D0252669

In Search of
The Canary Tree

In Search of

The Canary Tree

The Story *of a* Scientist, *a* Cypress, *and a* Changing World

LAUREN E. OAKES

Illustrations by Kate Cahill & Cartography by Erik Steiner

BASIC BOOKS
New York

Basic Books
Hachette Book Group
1290 Avenue of the Americas
New York, NY 10104
www.basicbooks.com

Printed in the United States of America

First Edition: November 2018

Published by Basic Books, an imprint of Perseus Books, LLC, a subsidiary of Hachette Book Group, Inc. The Basic Books name and logo is a trademark of the Hachette Book Group.

The Hachette Speakers Bureau provides a wide range of authors for speaking events. To find out more, go to www.hachettespeakersbureau.com or call (866) 376-6591.

The publisher is not responsible for websites (or their content) that are not owned by the publisher.

Print book interior design by Trish Wilkinson.

Library of Congress Cataloging-in-Publication Data
Names: Oakes, Lauren, author.
Title: In search of the canary tree: the story of a scientist, a cypress, and a changing world / Lauren E. Oakes; illustrations by Kate Cahill & cartography by Erik Steiner.
Description: First edition. | New York, NY: Basic Books, Hachette Book Group, 2018. | Includes bibliographical references and index.
Identifiers: LCCN 2018032911| ISBN 9781541697126 (hardcover) | ISBN 9781541617421 (ebook)
Subjects: LCSH: Ecologists—Biography. | Oakes, Lauren.
Classification: LCC QH31.O25 O25 2018 | DDC 577.092—dc23

LC record available at https://lccn.loc.gov/2018032911

ISBNs: 978-1-5416-9712-6 (hardcover); 978-1-5416-1742-1 (ebook)

LSC-C

10 9 8 7 6 5 4 3 2 1

For John,
who fell in love with these forests
and spent his life immersed in them;
And you, little one,
I tried to write quickly for you.

A truly noble tree, . . . undoubtedly the best the country affords, and one of the most valuable to be found on the Pacific coast . . . [W]hen these have reached the age of several hundred years the down-trodden trunk, when cut into, will be found as fresh at the heart as when it fell.

—John Muir, 1882

Callitropsis nootkatensis.

Contents

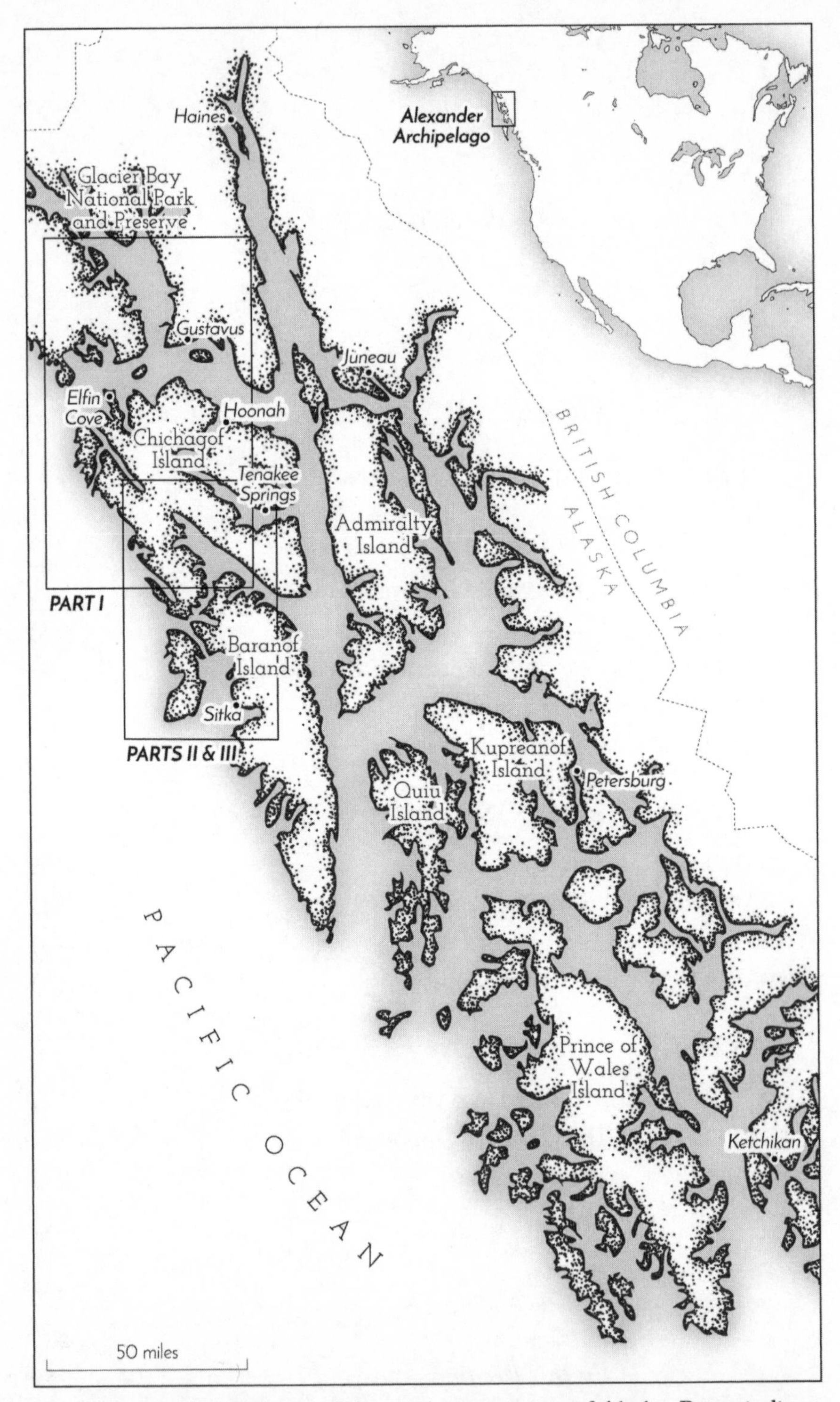

The Alexander Archipelago off the southeastern coast of Alaska. Boxes indicate general areas for research conducted in Parts I, II, and III.

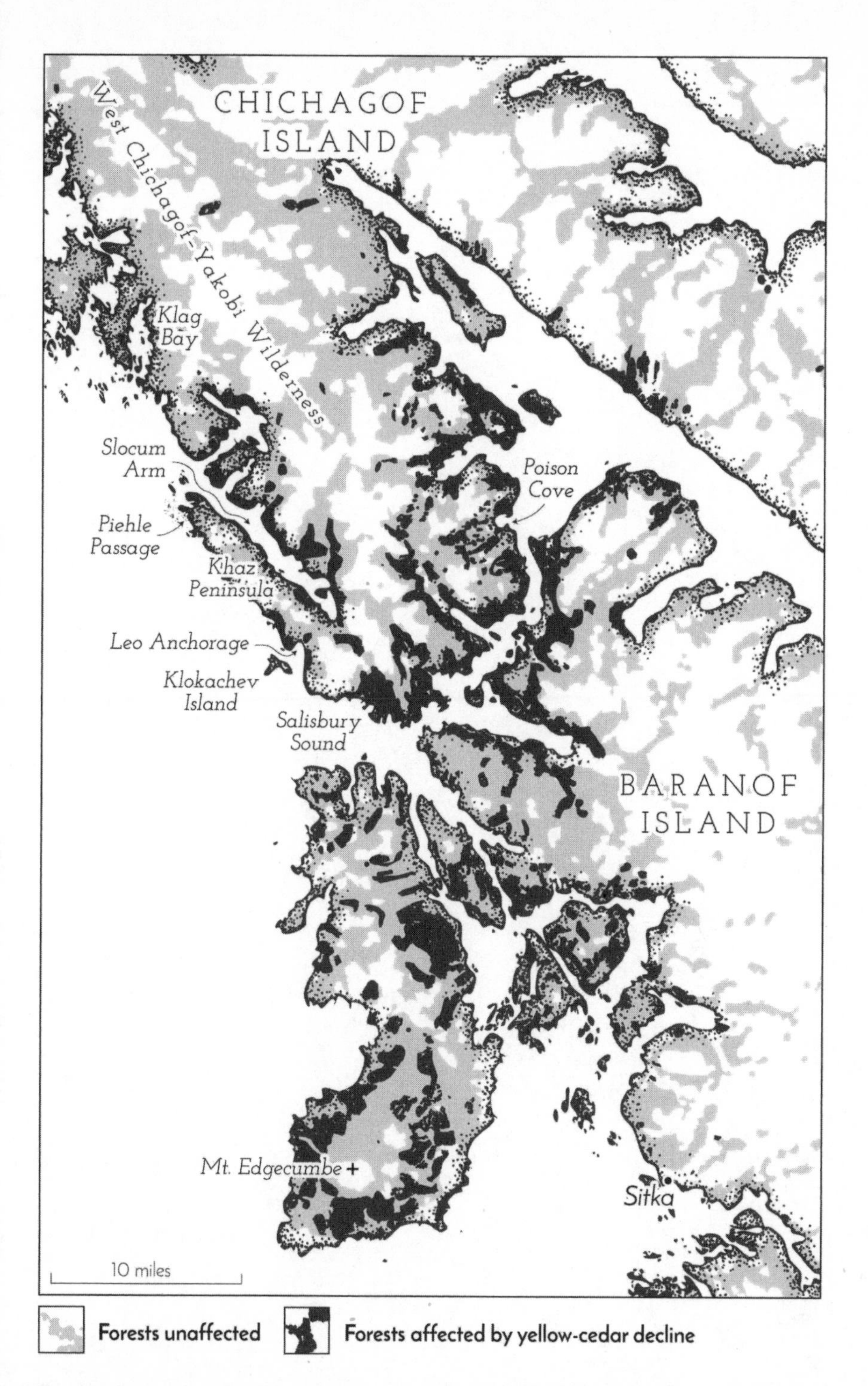

Detailed map for locations referenced in Part I. Light shading shows forests unaffected by the widespread yellow-cedar mortality. Dark shading reveals forests affected. (Data collected by the United States Forest Service.)

Prologue

I CAME TO Alaska looking for hope in a graveyard. Ice melting, seas rising, longer droughts—in a world seemingly on fire, I chose to put myself in some of the worst of it. The Alexander Archipelago in Southeast Alaska is a collection of thousands of islands in one of the scarce pockets remaining on this planet where thick moss blankets the forest floor and trees range from tiny seedlings to ancient giants. But I wasn't loading into that Cessna four-seater to look for fairy-tale forests of spruce, hemlock, and cedar. I was flying in search of the forests I'd study—the graveyards of standing dead trees and the plants I so wanted to believe could tell me, through science, that maybe the world is not coming to an end.

I tightened the seatbelt, tugged the straps on my orange lifejacket, and slipped the headphones over my ears to dampen the whir of the Cessna's engine. Avery, our pilot, flipped a few switches, and then the audio crackled into static.

"Let's check the sound," he said.

"Pretty clear," Paul reported from the copilot seat. "I can hear you."

"Got you both," Ashley confirmed to my left.

"Good to go," I said, swinging my camera off to one side so I could see the laminated maps on my lap.

I was five days away from spending the summer measuring dead and dying trees on the remote outer coast, where winds whip and whirl and ocean swells slam against the rocky shores. We were flying for science and flying for field logistics; I needed to finalize the practical stuff before it was really go time. The flight was supposed to confirm my sampling strategy for the forests with the two far more senior researchers present—Dr. Paul Hennon, a forest pathologist, and Dr. E. Ashley Steel, a statistician, both with the United States Forest Service. Any last decisions we made would determine the shape of my study—everything from which trees I included to the analyses I'd run years later. I could spend a summer on the coast and come back without enough data for statistical power. I could crunch my data for a year or two, and then realize I hadn't collected all the data I needed to answer the research questions, that something was missing.

The species is *Callitropsis nootkatensis.* (Botanists still argue over its genus, but more on that later.) Some people call it the Alaska cedar. Others call it the yellow cypress, or the Nootka cypress, named after Nootka Sound along Vancouver Island, where it was first botanically documented.[1] Alaskans use the name yellow-cedar, but really, a name is just a name. What mattered to me from the beginning was that these trees are long-lived, and that, though they are coveted for their golden wood, and culturally revered for their majestic and mysterious ways, they are dying in our warming world.

"Get ready to soak it in," Ashley said as the floats lifted from the water. She wiggled her knees in anticipation. The plane swooped around to head west, and I watched the bright blue glacier outside Juneau shrink into the distant landscape.

Basecamps, I thought. *Eyes open for good basecamps—coves far from creeks to avoid the bears; big, sloping beaches for boat and plane landings.* Safety was a legitimate concern out there. I could get stranded by a turn in weather, run out of food, or startle a bear, inadvertently inciting an attack.

"Think stratified random sample," Ashley said, snapping me back to scientific protocols.

Two years into graduate school, I was finally accustomed to translating science. Stratified—meaning I needed forest sites I could categorize into specific groups based on the presence and severity of the tree death. Random—meaning I needed a way to select sites that could give every patch of forest an equal shot at getting picked.

Patches, I thought, simplifying things. *We're looking for the large patches of dead trees.*

I looked down at the sea of whitecaps and wondered what we'd face in the months ahead. Then, after squinting at the green in an attempt to differentiate the tree species, I took the lens cap off my camera to snap a few photographs. I waited patiently and scanned the passing coastline for the cypress. Despite how much I had prepared, I really had no idea what I was getting myself into—scientifically, physically, or emotionally.

In the maps I held, I'd traced black lines around the dead trees from various datasets. The plan was to fly over them. I'd confirm they were there in reality (and not just in pixels), dead by the telltale signs—brown foliage and leafless branches, limbs lost to decay. Then eventually—one week, two weeks, or even seven weeks later—I'd reach a subset of them by foot and kayak with my crew to measure the dead and dying and whatever we found still green.

We had about thirty minutes until the plane reached Chichagof Island—the place where I really needed to pay attention. Testing my skills, I followed our flight path on the pilot's screen, trying to match our location with the maps on my lap and then verify with the view below. I managed to keep up for a solid ten minutes, not because I was skilled at the task, but because we flew mostly over water. Pretty soon I was totally lost, unable to reorient from the screen, the maps, or the view, and feeling nauseous.

I reached up and opened the small vent beside the window, angling the air toward me. Ashley's face looked pale.

"How ya feeling back there?" Avery asked through the audio crackle.

"Just fine. Little air seems nice," I said, hoping to avoid any mockery.

"Like vomiting," Ashley reported. "And I don't see any patches yet."

I closed my eyes and breathed deeply to calm the queasiness. When I finally opened them again, confident I wouldn't hurl, I didn't need to look at the maps or the screen to know where we were.

"Whooaaaaaaa," I exclaimed into the headset. We were flying across a long, narrow artery of salt water running into the forested land. "Peril Strait," I said.

"We're in the heart of it now," Paul declared. He sounded almost proud.

To the left, the verdant coastline broke off into inlets and side channels. To my right, I could see the steep hillsides covered in white skeletons of dead trees—standing on end like telephone poles, leafless ghosts of the towering cypress. Boulder-sized rocks on the beaches looked like little specks in relation to the large tracks of terrain with dying trees, the canopies of foliage in faded sepia tone.

I had been so focused on building a sound scientific study that wouldn't get me or my crew killed that I hadn't given much thought to what I would *feel* when I first saw the dead. From the bird's-eye view, the giant trunks looked like thousands of toothpicks stuck in the earth. If trees were people, anyone would have called it a tragedy—an epidemic running rampant throughout the community in the largest remaining coastal temperate rainforest on Earth.[2] I felt the tiny hairs on my forearms rise.

Sampling, I thought, attempting to return to the task at hand.

"Let's get a little more perspective," Paul said. Avery took us higher, enabling us to see farther inland.

As we cut across the island toward the coast, my stomach tightened, not from the nausea, but from the shock.

Random sampling?! The plan of picking random patches from the grid of latitudes and longitudes on my lap—a strategy that had seemed both feasible and logical back at the office—would never work. Even if I could manage to land my kayak at the closest beach, some of those forests—far from roads or even trails—would take days to reach. I simply didn't have that kind of time. Similarly, the idea of classifying patches of dead trees into categories of stressed trees, recently dead trees, or long-dead trees was ridiculous. No way could we get that level of detail from our aerial perspective.

Developing a new sampling plan at the last minute was potentially a huge setback, but I should have been more worried than I was. Instead,

I was distracted by a knot, deep inside my stomach, pulling tight around a sense of loss and fear. It was that kind of terrible feeling that surfaces when you drive past a wreck on the highway and the ambulances aren't racing anyone off to the hospital. You wonder who died, who's gone forever, what loves or lives are cracked. Who's left behind to begin anew? It could happen to any one of us. And then you drive on, because that's all you can do.

Except for me, for this, there was no driving on from the graveyards of standing dead, no going home, and no forgetting. I didn't know it then, but those trees would change my life. In the moment, soaring above them, they made me feel vulnerable to our warming world in a way I had never felt before.

There's a limit to the change we can tolerate, I thought. *There's a threshold and tipping point for every species—humans included.*

I said nothing. I looked at the screen to confirm that the inlet before us was the one where I'd spend the summer. Slocum Arm—another artery of ocean, this one sliced into the edge of Chichagof Island from the north.

Avery dropped down a bit closer. Even from the plane, the stark landscape felt eerie, as if the dead trees were signposts for an even greater tragedy to come.

"Pretty calm in here," Paul said with a reassuring tone. "This is about as protected as you can get from the open ocean." I relaxed a bit, noting a possible campsite off to my right.

"We have lots of dead trees," Ashley reported, "but I don't see any bands. It made a lot more sense in the office than it makes out here."

Before I ever set foot in the forests of Slocum Arm, so much had already gone wrong. But gazing down at the graveyards, I found myself thinking, *I'm not backing away from this.*

"I'll have to find a way to group the forests from the ground," I said, "when I get there."

Static hit in the headsets again.

"There's a way," Paul said. "Just not the one we thought." Paul later told me he was a bit worried that seeing the massive scale of the landscape and the patches would make me back out—just days before my real work would begin.

I agreed with Paul—I'd find another way to stratify the forests on the ground. I'd guided clients down white-water rivers, built trails in the Rocky Mountains, and backpacked alone in snowy winters. I could paddle. I wasn't afraid of the rain or the grizzlies (too much), and I trusted I could resolve whatever scientific challenges arose with my team. I had volunteers packing food in metal, bear-proof boxes. I had a boat captain in Sitka lined up to take us out to the coast. I'd already spent thousands of dollars in grant money. How I'd stratify the forests was only the first of many problems we would have to solve in the years to come. I'd find my way through the obstacles to understand the impacts of these dying trees on the surrounding communities of plants and people, to uncover whether this species was the canary in the coal mine—calling out for our own inevitable demise.

What I didn't know then was that these dead trees would eventually give me more than just hope. They'd give me a sense of conviction about our ability to cope with climate change. They'd motivate me to do my part. They'd move me from pessimism about the outlook of our world to optimism about all we still can do.

As we made our way back down Slocum Arm, I stopped focusing on the dead trees and started looking around them. I could see green peeking up and around the barren trunks. I wondered if there was a new forest forming and what individuals could survive amidst the changes occurring. They were there. I could see them reaching toward the light through the broken canopies. I was committed to finding an answer—but for more than just the fate of the trees.

PART I:
THE SLOW BURN

One way to open your eyes to unnoticed beauty is
to ask yourself,
"What if I had never seen this before?
What if I knew I would never see it again?"

—Rachel Carson

Introduction

On March 4, 2015, I stood before an audience of over a hundred colleagues, friends, and family members at Stanford University. It was my doctoral defense—the final hurdle to becoming a card-carrying scientist, to throwing *Ms.* and *Miss* (or *Mrs.* one day) out the door for *Dr.* I thought if I passed, an enormous weight would lift; I'd feel a great sense of freedom to launch into my scientific career. That wasn't how it went.

Long before our real, measured understanding of climate change began to emerge, the ecologist Aldo Leopold claimed that "one of the penalties of an ecological education is that one lives alone in a world of wounds."[1] But today, the stark headlines in the news are doing the same for a much broader public as they paint a picture of a frightening future: "Climate Chaos, Across the Map," "Greenland Lost a Staggering 1 Trillion Tons of Ice in Just Four Years," "Oceans Getting Hotter Than Anyone Realized," "Climate Change Is Killing Us Right Now."[2] But scientist or citizen, if you accept the reality of our current climate trajectory, I think we're all wondering if there's anything we can do and how best to live amidst this threatening sense of demise.

I had never given much thought to what spending six years studying the impact of climate change—on forests and on the people who depend on them—would mean for me personally. I'd never considered what it might be like to join the tiny pool of highly trained scientists living in that world of wounds. And what it would take to find a way forward.

It all started in 2010 with what I thought was a simple ecological question: How does the forest develop after the yellow-cedar trees die? While some of my colleagues outlined plans to project climate scenarios or predict drought—topics I consider among the hardest to pursue scientifically, and among the darkest to study in terms of their potential impacts for humanity—I actually thought my question was fairly optimistic. I wanted to know what species could still thrive amidst loss and change; what life could tolerate the conditions we're creating, and how and why. A question that began on an isolated coast later unfolded in communities scattered throughout Southeast Alaska. By kayak and foot, I traveled for months to forested sites scattered over miles of ragged coastline. My quest brought me into the off-grid homes of hunters, naturalists, and Native weavers, and into the offices of forest managers who were once responsible for clear-cutting old growth in America's largest National Forest. I was looking to document impacts, but I was also seeking solutions to a seemingly intractable problem.

When I'd entered graduate school, I'd never expected to find myself obsessing over a single conifer species, let alone one in the Alexander Archipelago. In fact, I'd always thought scientists who devoted years of their lives to just one species were pretty weird (so perhaps I was always destined to join their numbers). But don't get me wrong. If you're going to obsess over a tree species, *Callitropsis nootkatensis* is an awesome one to pick.

Majestic in form, monumental in size, *C. nootkatensis* is related to other cypresses, such as the giant sequoia of the western Sierras, and the towering alerce in the Chilean coastal range.[3] It's not a true cedar, like *Cedrus* trees in the pine family. But distributed in sparse patches and pockets, the yellow-cedar has persevered through centuries, even millennia, of change in the Pacific Northwest. No one knows exactly how. Marked by rings of growth, the record of every individual tells a complex story of good and bad years. The core of a yellow-cedar tree registers a long history of life events that the human eye can behold but only science can decipher.

In 1879, when the naturalist John Muir traveled from California to Alaska, he sketched yellow-cedar trees in his journals and described their branches as "feathery, dividing into beautiful light green sprays."[4] Native people have cultivated intimate relationships with these trees for thousands of years, using its bark in their weavings, its wood for totem poles and canoe paddles. Today, they are among the most economically valuable trees in the Pacific Northwest.

As fascinating, beautiful, and useful as the yellow-cedar is, it became the center of my study for one reason: the species, which had survived so much, was dying in large swaths on the landscape.

The earliest known report of the dead patches comes from a hunter named Charles Sheldon, who, in 1909, noted them in swampy areas.[5] Toward the end of the twentieth century, scientists observed high rates of tree death, inciting concern among Alaskans.[6] At the time that I began my doctoral studies, a team of researchers led by Dr. Paul Hennon had recently uncovered climate change as the culprit.

Places close to our poles are getting hotter quicker.[7] Temperature increases in Alaska have doubled the global average since the mid-twentieth century.[8] Warming—and all its consequences—are already lived experiences for Alaskans in reality, today. So focusing on the yellow-cedar for the yellow-cedar alone wasn't my drive. Simple curiosity about biological processes and species evolution wasn't enough for me. I wanted to do more than discover. Like most environmental scientists, I wanted to solve problems. I thought perhaps the ways in which Alaskans were coping with their changing environment and the loss of this majestic tree might offer a glimpse into my own future—into all our futures—as the effects of climate change continue to cascade across the planet.

CHANGING LANDSCAPES HAVE long intrigued and concerned me. As a young girl, I liked framing complicated relationships between people and the natural world through a camera's lens. My father gave me my grandfather's Kodak Retina camera, a relic of the 1950s, when I was fourteen. I filled my earliest contact sheets with images of an old maple tree in my backyard, documenting the scars of the many trimmings that had reshaped its ragged limbs over time. Later, with my father's 35mm Olympus, I photographed roads cut through fields, shrubs pruned in cultivated gardens, and urban sidewalks planted with spindly trees in squares of exposed soil. I was fascinated by the ways in which people shape the natural world, and that fascination led me from one altered environment to another throughout my twenties.

I traced sources of water pollution through flooding streets in Rhode Island. I witnessed communities and desert landscapes transformed by oil and gas development in the American West. I confronted mining development in pristine watersheds in Southwest Alaska, and road construction through the temperate forests of Chile. I worked as an environmental advocate, a documentarian, and a policy researcher. But I decided to become a scientist to learn how to assess the consequences of environmental changes more precisely and systematically. In retrospect, that maple tree in my backyard was probably the harbinger for the cypress that unexpectedly seized my life as a young scientist.

When I traveled north to Alaska, the biggest questions on my mind were about despair and hope. Should we all throw our arms up in the air and admit defeat? Is there anything that any one of us can do that will actually make any difference? What will make a difference? More and more, people are grappling with these questions as awareness of climate change increases. The big picture of the warming world feels overwhelming, unwieldy, and out of our control. Temperature projections show a planet turning red over the course of decades to come. Most scientists today show graphs and numbers, complicated models, and statistics that basically say, "We are too late." Even if we stopped all emissions and halted the pace of the life we've created, we are just too late. The trajectory still follows the arc of a wave until it crashes down on us.

But from 2010 to 2015—the years leading up to my defense at Stanford—I lived in a world of cautious hope. I studied what happens to other plants in the forest community after the yellow-cedar trees die, and how Alaskans were adapting to the changes in their local environment. On the outer coast, I encountered many dead trees, but even in the forests affected by the dieback, I found survivors. I wondered what had enabled those particular yellow-cedar trees to live on, and what had allowed other species to take over. Numbers and observations led me to some answers. Other truths came only from the people who knew the forests best.

My research was a blend of ecology and social science. Talking to people was just as important to me as measuring plants and recording temperature data. Sure, I formulated hypotheses and sought answers through systematic methods like all my colleagues were doing, but as a human being living in a world that faces all kinds of threats from climate change, I also searched for a way out of my own sense of fear and helplessness. I haven't talked much about that part—until now. As scientists, we are taught to be impartial and precise, and to avoid, at all cost, the personal. That goes in the black box, along with all the intricacies of the scientific process itself.

I did, in fact, make scientific discoveries. From thousands of plant measurements along the rugged outer coast, I found forests flourishing again. From hours of interviews with Alaskans who value this tree, I found a community of people developing new relationships with the emerging environment. On the day of my defense, I presented tables and graphs detailing all the ways in which people were responding to the death of the tree and coping with change. They had found substitutions for the yellow-cedar, and had developed ways of using dead trees. They'd sought opportunity, recovery, and innovation in the face of defeat.

My research had been published in one scientific journal already.[9] Other articles would soon follow.[10] But something was missing. In crafting concise, scientific language, I had stripped away the humanity.

"We measured 2,064 trees and 882 saplings"—in five words and two numbers I removed months of my own experience—living and feeling, listening and breathing, amidst those dead trees. In distilling 1,500 pages of interview transcripts into a single, elegant table, I'd left out the way a logger runs his calloused hand across fine-grained wood in admiration, or the silence that fills the room before an Alaskan describes an impressive yellow-cedar. I'd cut the stories from naturalists like Greg Streveler, or Tlingit Natives like Teri Rofkar, or loggers like Wes Tyler—the people I'd met who had found ways to benefit from the emerging environment while losing a species they used, valued, and loved.

When I'd asked people to describe a yellow-cedar tree, these were some of the words that broke the silence of reverence: sweet-smelling, rare, beautiful, alluring, breathtaking, strong, sturdy, sensual, mysterious, wise. One Tlingit Native explained to me how building a close relationship with nature, the kind most people only cultivate with another human being, enables one to cope with change. There wasn't space for those details in science, so I translated her eloquent words into data points. I remained true to what she shared in my analyses, objective in my interpretation, but the scientific process buried the essence.

The PhD "defense" was something I knew little about until I was required to do one. The process varies across institutions and countries, but the general gist is the same: a young scientist publicly presents his or her research, fields general inquiries, then proceeds to a private question-and-answer session with senior scientists who will judge the work's merit. At the end of it all, I paced the hallways alone outside that room, waiting for the verdict from the deliberation.

When the committee invited me back and referred to me as Dr. Oakes for the first time, I was surprised I didn't feel closure. I felt relief, yes, but not the closure I'd expected. Instead, I had this gnawing, urgent sense that there was more to be done. Something felt unresolved, and it was far more personal than scientific.

I had pushed hard to earn a place among high-level scientists, but I needed my work to exist outside of a professional echo chamber as well. Despite having written 223 pages on my research—which was soon to be signed, sealed, and delivered—I knew I was about to write as many for this book. I returned to my office not long after the defense with a box of journals and papers and started digitizing years of notes to resurrect the buried essence, to figure out, for myself, how best to live in the rapidly changing world today with what I'd learned.

Herein lies the story that felt untold.

This book is about a species—a tree called *Callitropsis nootkatensis*, how I fell under its spell, and how it inspired my search for people and plants thriving amidst change. It chronicles my effort to answer what happens in the wake of yellow-cedar death, not only to uncover the future of these old-growth forests, but to share lessons that apply to people on other parts of the planet. It is a book about finding faith, not of any religious variety, but as a force that summons local solutions to a global problem, that helps me live joyfully and choose what matters most in seemingly dark times. If we start looking at the local picture and the various ways in which we all depend on nature every day, solutions emerge. I witnessed this in Alaska.

The camera I carried on the outer coast to photograph the ancient trees looks nothing like my grandfather's old Kodak, which rests, unused for many years, inside a plastic storage bin. Megabyte files and thumbnails have replaced film and contact sheets. I am still focused on trees and landscape changes, but I have changed as well. We create and re-create narratives throughout our lives to make sense of what happened, to process experience, to interpret and reinterpret our view of the world as life unfolds. I believe that beautiful and difficult process is what it is to be human. So although the research I conducted shapes the narrative of this book, the writing and reporting in subsequent years were also part of the journey. The scientist I am today influences how I explain what I knew (and didn't know) back when my research began, as well as the details I've added since. How I personally responded to the challenges my work presented, and what I learned from the many people I interviewed, directed the most poignant events and conversations I chose to portray.

The writer, environmentalist, and historian Wallace Stegner once wrote, "If art is a by-product of living, and I believe it is, then I want my own efforts to stay as close to earth and human experience as possible—and the only earth I know is the one I have lived on, the only human experience I am at all sure of is my own."[11] In bouncing between

California and the Alexander Archipelago for all those years, I focused, scientifically, as intensely as I could, on this rapidly changing world, then, emotionally, on the struggles that experience triggered. I discovered parallels between the scientific and the personal as I confronted loss in my own life and found ways to move forward.

Scientific facts rely upon assumptions; they are blocks built upon one another. But what I learned in the archipelago came from a mix of science and the act of doing that work; of striving for another layer of understanding in lived experience. Our own truths, felt in the heart and known in the mind, are transient as we create the storied landscapes of our lives, again and again and again. So this is me, at this point in time, finding my way into tomorrow in a world destined, as some argue, to become uninhabitable.[12] It is a story of refusing my own fear of what a warming world will mean for me in my lifetime; a story of becoming an unexpected optimist against a backdrop of dying forests and in a profession where pessimism is often the common response.

A FEW DETAILS—

This is a work of nonfiction. The people, places, and events are real, and I came to them through my research in Southeast Alaska and at Stanford University. I haven't changed any names, except that I gave nicknames to every Paul that came after Dr. Paul Hennon—the lead scientist, who had linked climate change to the dying trees. Some, like Paul "P-Fisch" Fischer, officially received nicknames during our work together. Other Pauls received nicknames only in my head while I was writing this book. Forest pathologist Dr. Paul Hennon, field technician Paul Fischer, forest ecologist Dr. Paul Alaback, plant physiologist Dr. Paul Schaberg, bear-hunting guide Paul Johnson—what are the odds of so many Pauls in a forest so few people know? Surely, you would stop reading in a fit of confusion if I called them all Paul.

Many of the people I met and worked with as a young scientist likely never imagined that I would write more than academic papers, or that their names and our professional experiences together would become public. I've done my best to portray them and the events accurately, writing from thousands of pages of field notes, transcripts, research papers, email records, letters, and journals I kept and others my field technicians shared with me. Often, scientists told me about study results long before they were published. So, in some instances, I describe their research at the time—true to the chronology of the events—that ultimately preceded publication dates. Some conversations I reconstructed from my notes and memory, then fact-checked them as a reporter would do. When possible, I went back to the people who were there to reconcile perspectives in an effort to recount conversations and describe what happened as accurately as possible.

The forty-five Alaskans I interviewed for my doctoral research consented to being part of a scientific study, knowing the perspectives they shared would be reported anonymously as data points and excerpts and one day, possibly, published in a book—and that they would be fully attributed and identifiable. I thank those people for their trust and hospitality and for sharing a window into their lives. Some opened their offices to me. Others opened their homes and offered me a place to stay in remote communities. Some I waited patiently for days to meet. Others I spent days with.

From the conversations I recorded and transcribed, I've edited quotations for length, and occasionally for clarity. Given the extensive length of each formal interview, condensing conversations in the service of the narrative was unavoidable. I conducted those interviews with the full approval of Stanford University's Institutional Review Board (IRB), which oversees a review process designed to protect the rights and welfare of people in any research study. The IRB uses the term "human subjects," but I never thought of anyone I interviewed as a subject. I approached every person partly as a scientist and partly as a concerned citizen on a sojourn, seeking sage advice from someone who might just have a solution to a wicked problem. Did I find, in their words and in my research, the one, big solution to climate change? I did not. But I found something that will help get us there and enable each one of us to live more purposely in these times.

CHAPTER 1

Ghosts and Graveyards

I FIRST HEARD about the dying yellow-cedar trees nearly thirty years after the first scientific attempt to discover what was killing them. I was standing in the pouring rain with John Caouette, a forest statistician, outside the Paradise Café in downtown Juneau. Rain dripped over the hood of his bright red jacket. John hardly seemed to notice that we were both already soaked, and he was far more interested in talking about forests than seeking refuge inside.

"I found a place I loved," he said. "I've watched people come and go, but I never wasted any of my time wandering. I just stayed."

John was known among researchers in Southeast Alaska for changing the way the Forest Service assessed its stands of trees. He'd figured out in the late 1990s that volume, a simple measurement used for decades of planning timber harvests, failed to capture the true character of a forest.[1] A matchstick forest packed with slender trees could generate the same number of board feet for lumber as a grove with widely spaced giants. Mathematically, the amount of wood was the same, but ecologically, the differences in structure made for distinct communities of plants and wildlife. Using volume to inform management decisions overlooked the other characteristics that could make an old-growth forest home for black-tailed deer, the coastal wolves that prey upon them, and nesting birds.

Rivulets of water slid down John's jacket, slowly saturating his cotton pants and turning them from blue to darker blue. Across the street on the dock, tourists wearing yellow plastic ponchos emptied out of a cruise ship bigger than any building surrounding us.

"People spend years dreaming of a trip to Alaska," he said, nodding toward the crowd forming. "You know where we are?! We're in the heart of the Tongass—the largest National Forest in the United States!"

At seventeen million acres of land, the Tongass covers about 80 percent of the Alexander Archipelago. Boxed in by National Forest, ice fields, and ocean, Juneau, the capital city of Alaska, is accessible only by boat or plane. The population hovers around thirty-one thousand people. One main road, commonly called "The Road" by residents, runs thirty-nine miles along the coast. It ends, on both sides, against thick forest—ferns, blueberry brush, spruce, and hemlock trees.

It was late June in 2010, and I had arrived in Southeast Alaska on the MV *Columbia*, a ferry coming up from Bellingham. With Stanford's approval for "exploratory research," the summer's goal was to find a topic for my dissertation. I didn't want to focus on refining our climate predictions, uncovering crucial differences between future changes of a tenth of a degree or two-tenths. Many of my peers were eager to study the consequences of various climate scenarios. But to me, the more interesting questions were about the present, the current effects of a changing climate—and they pointed me to the north.

My parents aren't scientists; I don't know of any blood relatives who also earned a PhD. My mom taught high school chemistry for ten years and then worked in elementary school administration. Her father emigrated from Italy through Ellis Island at age eleven. He was alone and couldn't speak a word of English at the time, but later became a medical doctor at Yale–New Haven Hospital. He died before I was born, but the stories my mother told me portrayed a persistent man who created a life centered on compassion for others. My own father was the only child of a middle-class family in Ohio. Recruited from the land of cornfields to Harvard in 1963, he was the first in his immediate family to get a college education. As a little girl, I watched him strive for success in the corporate world, then set out on his own as an entrepreneur; he wanted to start businesses that would endure. Some did well; some did not. Stress in my family came, in part, from the lack of security in his work.

Others in our East Coast community led more affluent lives. Their values seemed at odds with the ones my parents held for less materialistic living. New cars raced around six-lane highways, but my father kept his beloved 1979 classic running for years. I thought I could help alleviate the undercurrent of financial strain in my family, which I felt keenly at times—especially in later years—by wanting and needing less. I grew up in a saltbox home built in 1774, the last house on the border between Connecticut and New York, and I watched modern homes replace the green fields where I played down the street. I was acutely aware of the growth and consumerism I was witnessing in my hometown, which had disproportionate effects I couldn't yet grasp.

By the time I went to college, I was into the environmental classics—Thoreau's *Walden, or Life in the Woods*, Edward Abbey's *The Monkey Wrench Gang*, Rachel Carson's *Silent Spring*. I viewed Abbey's attack on the expansion of dams, and Carson's efforts to halt the use of pesticides, as fights to preserve human rights. If air and water were basic human needs, then so were the wild, functioning ecosystems that perpetuated their existence. To me, even back then, taking care of human health and environmental health were one and the same. I toyed, briefly, with the idea of becoming a medical doctor like my grandfather, but Abbey and other environmental writers, such as Terry Tempest Williams, drew my attention to the American West. After nearly thirty-five years of marriage, my parents divorced a few months before I graduated. They remained close, caring for one another in a new, changed relationship. I moved west. Ultimately, I landed in science by following an innate concern for the ways in which people's actions affect the natural world, and how those changes might circle back—with positive and negative effects—on people.

My meeting with John Caouette was one of many I'd scheduled for Juneau and farther north. I planned to travel from the archipelago up to the Arctic in search of a place, or a community, or an issue—such as permafrost melting, or sea-level rising—that I thought might offer new insight into the current consequences of our changing climate. I never expected a cypress tree to be the hook.

My legs were getting cold outside the café, and I could feel water soaking through the cuffs of my jacket.

"Most scientists are using climate to predict where the trees will be," John said, swaying in his sopping sneakers. "But I'm using the trees to show what's happening with the climate—the other way around." John said his approach was unconventional and that scientists he was working with contested the idea, but he fundamentally believed trees would reveal something weather stations could not. I immediately liked the idea that an unconventional research approach could uncover new knowledge from the trees.

Once inside the café, I took notes in my black Moleskine notebook.

"Out of the six major conifers here in Southeast Alaska," John said, "yellow-cedar is the most sensitive to climate change. But it's such a curious and elusive species. Why do we find yellow-cedar at Shelter Island, just miles away, but only in a few sparse patches here in Juneau?"

John explained that no one really knew why it grows exactly where it does, or how much remained following the pulse of logging that swept through the archipelago from the 1950s to the 1990s. He said yellow-cedar trees have a certain "allure"—that Natives in the archipelago and farther south in Vancouver and the Queen Charlotte Islands in British Columbia have long revered them.

"The tightly grained wood is just beautiful—high value, really. And in comparison to other conifer species in the forest, yellow-cedars are relatively rare." He estimated that roughly 10 percent of the conifers across the archipelago were yellow-cedars. In some places, large cedars cluster on craggy slopes, but in the soggy soils interspersed between jagged terrain once carved by ice, they grow densely in conditions other species find less favorable.[2]

The archipelago ecosystem itself was intriguing to me already. Often overlooked because of popular interest in the tropical forests that are pervasive around the Earth's equator, coastal temperate rainforests are also relatively rare. They typically occur at the dramatic intersection of mountains and ocean, where rain falls heavily. Like a mirror of Southeast Alaska, a narrow strip of temperate forest runs down the coast of southern Chile. Other temperate rainforests are found in Tasmania, Argentina, New Zealand, Japan, and where Turkey and Georgia cradle the Black Sea. Collectively, they cover only about 1 percent of the Earth's land surface.[3] John said there was high commercial demand for yellow-cedar locally as well as internationally. While colleagues in my research group at Stanford turned to forests threatened by agricultural development in the Congo, Indonesia, and the Brazilian Amazon, I was facing a different challenge: a relatively rare species, of notably high economic value, in a relatively rare forest, at risk from a very different kind of stressor.

"If you take the ferry to Sitka you'll see dead trees all along Peril Strait," John said. "Put climate change on top of an overharvested species of cultural significance, and you've got an interesting situation."

I agreed.

"You need to meet Paul Hennon. He's out the road in Auke Bay with the Forest Service." I wrote Paul's name in my book and drew a rectangle around it.

"That man can pretty much tell you everything we know about yellow-cedar," he added. He gave me more names—Ashley Steel, the forest statistician, and about a dozen others. I started a list of what I would later call the cedar circle. Bigger than Texas, California, and Montana combined, Alaska is enormous, but the population is smaller than the city of San Francisco. Eventually, if you spend enough time focused on a topic like forests or fisheries, you get to know everyone else connected. John already knew them all. Eventually I would, too.

John scoffed at my plans to travel farther north to the Arctic. "Why would anyone ever go anywhere else?!" he asked before we parted, encouraging me to turn my scientific attention to the archipelago. "Go see Paul, and let me know what happens."

In the downtown public library, my first Internet searches on Dr. Paul Hennon revealed over a dozen published studies with titles pointing to old-growth forests, yellow-cedar, climate, snow, and disease. Trained as a forest pathologist—a tree doctor with expertise in identifying the fungi, viruses, and other maladies that attack forests—Dr. Hennon had spent a large chunk of his life studying *Callitropsis nootkatensis*. A news article published in the local paper in 2006 described the graveyards John had referenced and I had yet to see. The reporter, Elizabeth Bluemink, wrote, "Along the shores of Baranof and Chichagof islands—the heart of the species' range in the Panhandle—dead yellow cedars' bare gray trunks stick out like ghostly spines. Similar dead patches also have been noted in British Columbia."[4] Forest Service researchers had mapped about five hundred thousand acres of dying trees, Bluemink reported, and they were gathering evidence that suggested climate change was the culprit. At the time, Hennon and his team were analyzing new data that pointed toward the loss of snow as one of the key factors in tree death.

I picked at words to craft an inquiry to Dr. Hennon. I spent far too much time obsessing over my first impression and finally pressed Send, hoping I'd hear something, anything. He wrote back within minutes—a short informal reply with open arms: "thurs (tomorrow) would be okay for me. better than friday . . . so yes, please come see me."

I still didn't know what I was chasing and why.

The eight-mile road from downtown Juneau to Auke Bay is a straight shot along the coast—Gastineau Channel on the left, Douglas Island in the distance, open valley on the right, and then ridgelines running up into the Tongass to forest and rock face. Westerly winds blew steadily Thursday morning, and whitecaps surfaced on the channel beneath a pancake of gray cloud. Looking westward, I could faintly see the Chilkat Mountains breaking through the fog—craggy peaks glowing in soft pink light.

Inside a cement building guarded by a bent chain-link fence, the receptionist at PNW, the Pacific Northwest Research Station at the Forest Service, had me sign my name, fill in Dr. Hennon's, and add my entry time before directing me to his office.

Paul was sitting behind a heavy hardwood desk with his eyes locked on a computer screen. He had dark brown hair and a neatly trimmed salt-and-pepper beard. A bright yellow construction hat rested in the far corner of the room alongside a pile of gear—measuring tapes, a GPS unit, and a hand ax. Books with tan and green spines lined the walls. The room smelled like a freshly cut Christmas tree inside an old library.

"So, John Caouette has you thinking about cedars," he said, turning toward me, "and from what I gathered in your email, you're on the search for a research project."

Paul said that in his years of working for the Forest Service, he'd had few opportunities to work with young scientists; mentoring students through the scientific process was something he'd missed out on.

"Here's what I know from my own experience. You're on the cusp of choosing what you'll study for four or five years and maybe more. I was in your shoes in 1981, and a forest pathologist here suggested I try to identify what was killing the yellow-cedar trees." He took a bite of his sandwich, smiled, and then declared, "Now, almost thirty years of work later, we're close to publishing a synthesis on the answer."[5]

"Snow," I said. "Climate change. I read a few of your studies at the library yesterday."

"Yes," Paul agreed. "It's a complicated pathway with a number of contributing factors, but climate change plays a critical part."

Paul told me about a series of studies he had conducted with a scientist named Dr. Paul Schaberg, a plant physiologist out of Vermont (enter Paul #2). That work led to "the cliché smoking gun" for cedar. They'd subjected seedlings to cold weather snaps and used perlite, a mineral that acts as an insulator, to test how snow cover could protect roots from sudden temperature fluctuations. When they exposed the fine roots of yellow-cedar trees to cold events that dropped below –5°C, the roots responded unlike those of any other tree species in the archipelago's cold forests. They died. Snow acted as an insulator—a blanket critical for protecting the trees.[6] Loss of snow cover left the yellow-cedar trees vulnerable.

"So that's the counterintuitive part," I said. "Death by freezing in a warming world. The trees are actually dying when the roots freeze?"

"Here in Southeast Alaska, we're poised in a threshold environment," Paul explained. "Winter days often fluctuate between rain and snow. With warming temperatures, we're getting more precipitation as rain, and less as snow, but sudden springtime cold events still persist." The trees couldn't survive the cold snaps without the white blanket guarding their roots. Chuckling a bit, Paul called the three-part series of studies they conducted "The Schaberg Trilogy," as if to indicate that the killer they'd identified was straight out of a sci-fi mystery.

Then he'd shifted from labs to the landscape, mapping dead and living yellow-cedar trees on a dormant volcano, Mount Edgecumbe, near Sitka. Its upright, conical structure made for the perfect place to study where yellow-cedar trees died in relation to the snow line. What he found was that the trees were dying at low elevations, where what used to be snow now fell as rain.

When I asked Paul what made the yellow-cedar more susceptible to climate change than other species, he said the response was likely due, in part, to the colder conditions present when the trees standing today first established themselves long ago.

"We think yellow-cedar trees persevered in glacial refugia during the last major ice age that ended roughly twelve thousand years ago. Then they flourished under the cool, moist conditions that followed. When temperatures dropped again during a more recent period called the Little Ice Age, only hundreds of years ago, they colonized the coastal lands where snow persisted at the time."[7]

"The low elevations where we're losing snow now is where they're dying," I confirmed. Paul nodded.

I took notes on details to look up later, trying to spare him from having to explain more. Irrespective of climate *change*, the human-induced phenomenon we know today, forests have been adapting to changes in climate for hundreds of millions of years. So natural versus human-induced causes (what scientists call *anthropogenic*) can get confusing. Climate, historically, has never been constant; it's driven by a complicated brew of factors, including plate tectonics, the Earth's shifting orientation to the sun, even asteroids hitting the planet. Through warming and cooling, some species have evolved, while others have gone extinct. Trees have inched up mountainsides seeking a cooler climate and then crawled back down.[8] When glaciers advanced, individuals survived in ice-free pockets—the refugia Paul had mentioned—as refugees of an ecosystem in flux. As ice retreated, they slowly expanded in number, making the most of the freshly exposed habitat.

The deeper in time we go—from tens or hundreds of thousands of years to hundreds of millions of years of plant life—the less we really know. But on the order of centuries, or a millennium, some of the oldest living yellow-cedar trees survive, today, in conditions known to be entirely different from those they encountered at the beginning of their lives.

Their death, Paul explained, also has to do with the timing of *de-hardening*—a process that trees growing in places with cold seasons go through every spring when the temperature warms. Trees prepare for winter by developing a solute that—like antifreeze in a car—protects the roots from freezing. With spring warming happening earlier and earlier, the trees drain the pipes too soon. Without the antifreeze and insulation, they're vulnerable to bitter cold snaps that break free from the high-pressure system in British Columbia.

"Most scientists are still trying to understand how climate change may affect various species," I said. "You've made the link already."

Paul's eyes avoided mine for a moment and he shuffled in his chair. "It was a team effort. There's still more we have to learn."

He got up from his desk and walked over to a map Scotch-taped to the side of a filing cabinet. The numbers 55, 56, 57, and 58 marked the arcs of latitude across the region. In just three hundred miles from north to south, nearly twenty thousand miles of crenulated coastline wrap the islands, but the map reflected a simplified archipelago with large masses of land fragmented by straits and inlets. Paul's map had square grids printed across the region and dots strewn along the coast and inland.

"Each dot is a place where we've collected foliage samples," Paul said. "We use a shotgun and shoot branches up high until sprigs of live yellow-cedar fall below. It's hard to distinguish yellow-cedar pollen from juniper and western red-cedar pollen, so we're using genetics instead to study how the species has migrated since the last ice age. We're also still trying to find ways to identify yellow-cedar trees by satellite imagery; that would give us a real assessment of what's left."

None of Paul's ideas sparked any interest for me. I couldn't imagine blasting branches with a shotgun, spending years looking at pixelated forests on screens, or pipetting samples wearing rubber gloves and goggles in a laboratory. It all seemed like fiddling while Rome burned—playing idly in the slow, steady burn of our warming world while perhaps other scientific questions could lead to solutions.

Paul returned to his computer, sifted through a couple of folders, and then opened a series of images of dead trees.

"What will those forests become?" I asked.

"We don't know. It's taken so long to figure out what was killing them. That's where the scientific interest went first." He smiled. "I like what you're thinking."

Before I left, Paul walked me outside and around the back of the building. There was a row of black plastic pots with seedlings growing—some no more than a foot tall, others beginning to look too big to be confined. The seeds had come from various locations throughout the archipelago. Paul and his colleagues would later plant them in different places to see if individuals with different genetics could resist freezing.

I walked along the line and stopped at the last pot, then reached out and touched one delicate branch. Up close, the foliage looked scaly, but it felt soft resting flat in the palm of my hand.

I inhaled the sweet smell of yellow-cedar.

What does this tree have to teach us? I thought.

Yellow-cedar foliage.

CHAPTER 2

Stand Still

IT FELT RASH to abandon all the exploratory research plans I'd put in place from California for the first idea I came across. So I didn't. I'd arranged to talk with government officials, resource managers from fisheries and water resource agencies, and scientists studying various problems—such as forest fires and melting permafrost—that had been highlighted in a recent state assessment of climate impacts.[1]

Before I left Juneau, I took a day off from meetings and volunteered at a local BioBlitz, a community effort to survey species of plants, birds, insects, and more in a designated area. It was there that I met Jonathan—a tall, friendly, banjo-playing bird biologist. Like me, he was splitting his time between Juneau and the California Bay Area. I fell for him fast.

In the weeks that followed my meetings with Paul and John, I was only half-present in many of the conversations—not because of Jonathan, but because of the yellow-cedar. Frozen Arctic soils store massive amounts of carbon, and what happens to that carbon as warming occurs plays a huge role in shaping current and future climate patterns.[2] If you're a young scientist and you're interested in climate change, working in the Arctic is about as sexy as it gets. But as much as I tried to get excited about studying fish showing up in new rivers as stream temperatures changed, or Native hunting grounds shifting as wildlife populations sought more suitable habitat, or even human communities on the coast farther north, facing the risk of falling into the sea, I couldn't.

Even as I found myself in Anchorage and Fairbanks talking to scientists who were studying recently burned forests, and then farther north, by the Beaufort Sea, with ecologists who were focusing on methane seeps from the frozen soils that were collapsing into thaw lakes, I was stuck on a cypress species over a thousand miles away. Sometime around August, I stopped forcing myself to stay open to other topics and let all my attention go back to the archipelago. I stopped trying to sort out what was so intriguing about this dying tree and accepted it, instead, as just that.

I BEGAN BRAINSTORMING research questions as I was trained to do. How have the forests of Southeast Alaska changed over time? How have communities adapted? Still in Fairbanks, I'd awaken in the middle of the night when the t-shirt tied around my head to block the midnight sun inevitably slipped free, and then I'd roll over to write. It took time to translate my interest and intuition into scientific questions. A concerned citizen might ask, "How big are these graveyards?," for example, but a scientist asks, "What is the spatial extent of forest mortality?" I wanted to know about the green peering up and around the dead trees, but an ecologist writes, "How is the species composition changing?" A concerned citizen might ask, "Does the death of this species affect people?," but an interdisciplinary scientist asks, "How are people responding to the changing ecological dynamics?" The beginnings of a study started to form in my mind: Across the archipelago, stands of yellow-cedar had been affected by the changing conditions at different times. What could I learn by comparing forests affected years ago with those experiencing widespread mortality more recently, and even with forests still unaffected? Would I find other species cropping up to take the place of cedars, or a forest community falling apart?

Eager to follow up in Juneau, I contacted Paul again.

"I've been reading through whatever I can find online," I wrote. "I'm curious if anyone has started looking at species composition across dead stands at various ages. I was thinking about how it might be interesting . . . to look at a few sites of stands that died at different time intervals. I am currently stuck on that idea."

Paul wrote everything without capitals and hardly any punctuation, but I liked his casual, confident style—"changes in stand composition, or plant succession," he noted, "is a high need area for our research program on cedar decline. with it we can interpret what happens next in the declining stands." Conditions, he said, have had some time to play out. Paul offered support from the lab in Juneau, help with logistics, which he affirmed were "not trivial," as well as assistance in the field and general guidance. I was encouraged.

I knew of "succession"—the word Paul had used—as one of the earliest concepts in ecology. Back at school, I'd read about Henry Chandler Cowles, an American botanist who had put the concept into practice with a paper in 1899 describing the way vegetation develops on sand dunes over time.[3] The central idea of succession is that an ecological community will follow some sort of predictable path after a disturbance or the creation of new habitat. Forests form as glaciers retreat, and regenerate after a fire. Plants colonize new ground after a landslide or lava flow. Ecologists have since debated various aspects of the pathway and called it an oversimplification of dynamics now known to be far more complex.[4] Whether I could find a predictable pathway or not, we didn't know. But Paul was using the ecological term to refer to what I was after—the question of what happens to the forests after the yellow-cedar trees die.

I canceled some meetings and scheduled others. I reoriented everything I had planned in my time remaining that summer for yellow-cedar and only yellow-cedar.

From a dorm room at the University of Alaska–Fairbanks, I made my way through the rest of the names John Caouette had given me. I looked up bios and backgrounds on the Internet and contacted some by email. I tracked down papers and news articles. I got around town by foot and public bus and eventually caught a train south to Anchorage. I needed to get a sense of what others were working on related to cedars, and whether the implications for the potential loss of the species were as pressing as Paul and John had suggested. I tracked down anyone I could who was somehow connected to snow, climate, or trees in the archipelago. I searched for relevant datasets, to no avail. It was becoming increasingly evident that I would need to get to the trees myself. I couldn't rely on any already-existing data the way many of my colleagues were doing at Stanford. I still had no idea where exactly I'd go to get my data, but I wrote Paul to tell him I was "sold on cedar."

I consolidated everything I'd learned about yellow-cedar and everything I wanted to know into a five-page memo that I sent back to my advisers at Stanford and to Paul. I wrote it as I was trained to do—with concise sections labeled Background, Purpose, Methodologies, Site Selection, and Research Timeline. For the specific methods I didn't know yet, I substituted various approaches that seemed plausible, assuming that only a handful would be the ones that worked in the end. I wrote a section for funding needs, and then a long list of topics I didn't know enough about and would need to learn. Although Stanford has a fabulous biology program, the learning model for graduate students is more about mentorship than curriculum, and formal coursework in field methods is sparse. I applied for approval—and received it—for a semester at the University of California–Berkeley, where I selected courses in the forestry program. I bought a ticket to go to Juneau and then another to return to California.

Back in the Auke Bay offices in Juneau, Paul passed me a map printed from Google Earth of the outer coast of the archipelago. I was looking at a large, pixelated landmass labeled Chichagof Island and a bit of mainland extending farther north. White covered the inland area. Ice fields and glaciers reached down into ocean bays. A long, narrow yellow box outlined the curving coastline from one end of the page to the other. Paul said he'd been thinking about my idea of studying forests affected by the dieback at different points in time.

"Here, at the base of Slocum Arm," he said, pointing to an inlet of water the shape of a cowboy boot, "we've mapped forests where trees died long ago. This spring—on that same survey I mentioned to you before—we mapped trees just north that looked stressed." He ran his finger along the coast inside the yellow box, upward from the arm. "Some recently dead, and some dying here."

"It's spreading? What about farther north?" I asked.

"There's yellow-cedar in Glacier Bay National Park and Preserve," Paul replied, pointing to the green coastline bordering the ice fields at the top of the page. "We don't actually know how much is there. Nobody has ever mapped cedar or estimated its abundance inside the park. But as far we know, it's there, and it's alive."

"We're talking remote," I said. "Like really remote."

"No roads. No trails. Probably no shelter. There's an old cabin from miners, but it collapsed long ago. You can only get to Slocum Arm by boat or plane, and not on any day. You'd need weather on your side."

"I could select sites with trees on Chichagof that died at different points in time, and then others inside the park where healthy trees remain, and then compare them."

"Exactly. The yellow box is what I was thinking for a transect—the outer coast transect." A transect is an artificial slice of land, a line drawn systematically across habitat as a way to bound research to a specific area.

"A chronosequence," I said.

He nodded.

A chronosequence is essentially a "space-for-time" substitution; space-for-time versus time-for-time is like the difference between comparing a young person to an old person and following one individual for decades.[5] Instead of studying how a single strand of trees changes over time, a scientist using a chronosequence selects sites that are similar in many ways, but differ in the amount of time that's passed since a given phenomenon has occurred. Time may have passed since a volcanic eruption, since a glacier retreat, or since wind events have moved sand to create dunes—as in Henry Cowles's study. But for me, the chronosequence would be about the time that had passed since the death of trees triggered by persistent cold snaps in a forest losing snow.

"The closest towns would be Sitka and Gustavus." I was jumping ahead. "I'd need a base to restock, dry out, and store gear."

"Field conditions on the outer coast would be challenging," Paul noted. "But let's address that later."

He directed me to a man named Greg Streveler, who'd surfaced numerous times already. For every name I'd crossed off in the days between Fairbanks and Juneau, I'd added another three or four, until the same ones kept coming up. Dr. Paul Alaback, the scientist who had first hired John for forest work years ago, had been the first to tell me about Streveler. I'd drawn stars around his name as Dr. Alaback had described his extensive knowledge of the forests, geology, and history of the archipelago. Paul said Streveler would know about yellow-cedar in the National Park and whether fieldwork on the coast was logistically feasible. It looked like I had one last trip to make before I returned to the six-lane highways of California. Jonathan and I went on hikes and ate fresh salmon. It wasn't goodbye when we parted in Juneau; we'd see each other down south in California.

ONE MAIN ROAD without clear lanes or lines runs from the airstrip in Gustavus through town, which is actually just an intersection called Four Corners. The road heads out past homes hidden among spruce trees to the edge of Glacier Bay National Park and Preserve, which is part of a twenty-five-million-acre World Heritage Site and one of the largest protected areas in the world. Unlike "The Road" in Juneau, "The Road" in "Gus" cuts across a long, flat plain through open fields and young forest. I walked from the road to Greg Streveler's home on fresh ground, land covered by sea only a couple of centuries ago. The earth in Gustavus is rising and has been for quite some time; as glaciers retreat, the weight of ice slipping away causes the ground to "rebound," like the soft surface of a bed springing back as a bowling ball rolls across the surface.[6]

Greg said he'd be home chopping wood for the winter, and I found him by following the echo of an ax down a gravel road off the main one. He was outside, swinging the blade beside a shed. His hair was white, his skin weathered with age, but his body was still strong and sinewy. He was wearing wool pants, with suspenders pulled up over a collared shirt. I felt like I'd come a long way to have a conversation with a legend. I was nervous, and I had no idea how it would go. I certainly didn't expect this man, who was so clearly revered by so many scientists for his depth of knowledge of the region, to tell me that science wouldn't give me all the answers, and then to become one of the most important mentors of my life.

Greg put down the ax when he saw me approaching and positioned a few more logs on the stack. He didn't bother with a self-introduction; nor did he seem to need verbal confirmation that I was Lauren from the phone call. He brushed his right hand on his pants and extended it toward me for a firm shake.

"I'll get you something clean to sit on," he declared and then disappeared to rifle around the shed.

"Oh, I don't need clean!" I said, not wanting to come across as high maintenance. I was dressed in the same dirty jeans I'd worn all summer, along with my rubber boots.

"Well, we should at least start you off that way," he replied, and then popped back out into the open with a five-gallon bucket in his hand. A crack ran down one side. He brushed some dirt off the bottom and placed it bottom up next to the hemlock stump where he was chopping. Greg took a seat on the stump and then gestured toward the bucket for me.

"That spot you're sitting on now was in a bog, a marshland, some one hundred years ago," he said. I hadn't told him much on the phone in our brief call—mainly that I was a graduate student at Stanford, I'd connected with Dr. Paul Hennon, and I was interested in studying what happens after yellow-cedar trees die. That question was pointing me toward a landscape he apparently knew quite well. Greg took the lead with an agenda crafted from that limited information, and I sat on the bucket feeling like an odd mix between a pupil and a subject of study myself. He walked me through his time along the outer coast in places like Dundas Bay, Lituya Bay, Graves Harbor, and other refugia near Icy Bay. I wrote down names—not of people this time, but of locations—where Greg had seen yellow-cedar, where ice had once been, where trees had survived in pockets of habitat after the last ice age. Like John and everyone else so far, Greg said that why the species regenerated well in some places and not in others was a mystery. He called the distribution of the majestic yellow-cedar "enigmatic."

When I could, I interjected questions about his background. He'd come first to Alaska as a college student, working for the university museum on an assignment in the Aleutian Islands. Later he stayed on at the University of Wisconsin–Madison to get his master's in biology with a minor in geology, while pursuing a project of his own back in Alaska. Greg said he'd made a decision to specialize geographically so that he could generalize topically. Over decades of research in various capacities, he had tried to apply whatever disciplines he could to "master one landscape."

"At some point," Greg explained, "I think science became a pastime to some. There's such a need to shape questions that have meaning, to create knowledge and understanding of a place over time. Now, I'm thinking more about meaning with a capital M."

He shifted to probing questions about where I'd grown up, how I'd landed at Stanford, and whether I actually thought we could combat the rapid rate of environmental change occurring with the slow rate of science. From all the research he'd done over the decades, he'd come to the conclusion that commitment to place was, in the end, more valuable than any "scientific" discoveries he made. Commitment to life and observation in one place led to discovery on its own.

"Your question—what happens after the trees die," he said. "So what? If you answer it, what then? On to the next project? Scientists have proven, again and again, a remarkable capacity for simply monitoring a species to extinction."

I squirmed inside, feeling pressured and judged, but held myself still on the bucket.

"I'm not interested in science as a game or pastime," I said. "I can assure you of that. We're both seeking meaning." I paused for a moment to meet him there. "I don't think we can stop the current trajectory," I continued. "So if warming is inevitable, I want to know what species will turn over, what communities will take shape, and, quite honestly, how I can confront something seemingly so massive and unwieldy in my own life."

"So, the yellow-cedar is your muse," he pressed.

"No," I rebutted. "I think it's a window into our future."

I took a deep breath, surprising myself with my own clarity. Greg smiled, and his rigid posture relaxed lightly on the stump.

"The world rewards scattered knowledge," he said. "It's much harder to choose a place to sink or swim, for whatever happens. That's going against the grain. What the world needs is more people with roots and the discipline to take the time to understand a place. There's a discipline required to resist the urge to scatter." He let that sit with me for a moment, and then continued.

"If you're choosing to focus on this place and committing long enough to answer what you've outlined, I'll help in any way I can. But know this: You won't be able to understand a place without standing still there. You can't pop in for only one snapshot. You need the moving picture frame. It is in the deep knowing of place, and in the act of observing changes over time, that allows us—even as scientists—to understand what we cannot measure."

I felt like he'd challenged me to a duel by probing both my integrity and my intentions, but, as I walked down the road back toward Four Corners, I realized it was a duel I'd already entered on my own. I didn't want to become an ecologist to monitor a living species to extinction, write it up for a journal, and then move on to the next study. I didn't want to crunch numbers and present findings from a "system" of plants and people I would never get to know. Only later would I realize how significant these ideas—of commitment to place, and of making my own observations in place—would become for me.

The values that stem from standing still and knowing a place would ultimately lead me to knocking on doors and asking strangers in the archipelago to talk with me about their relationship to the trees. I'd observe change occurring across time through the chronosequence on the outer coast. And this tension between scattering and standing still would gnaw at me personally—as I later struggled to puzzle out what actions matter most in the face of all we know today about our warming world.

I watched the Gustavus flats fade away beneath a thick cloud layer in the twenty-minute flight back to Juneau. When the Cessna dropped back through the gray, a sandy outwash terrain gave way to sheer mountains abutting ocean. I followed the line of The Road along the channel, and a set of photographs surfaced from my memory. Black and white, arranged in a four-by-four grid, the images showed a horse galloping. In the first photograph, the horse had only one foot on the ground, and in the second all four were in the air. They'd been taken at Stanford's Palo Alto Stock Farm—now the campus of Stanford University—by a man named Eadweard Muybridge, in 1878. He'd been hired by Leland Stanford, a former governor of California, to resolve a debate over whether all four feet of a running horse left the ground at the same time. Today, the photographs, now widely reproduced, are known best as the precursors to motion-picture projection. The only way Muybridge could find an answer was by capturing multiple perspectives of the same phenomenon—motion and change through stillness. Flying over a verdant landscape once covered by glaciers with the photographs of the horse in my head, I thought about my own chronosequence as a series of snapshots in time. When taken together, they might reveal the patterns of an otherwise elusive future. The chronosequence would be my moving picture frame—my answer to what happens next. Does the horse's foot ever leave the ground? Will climate change reshape these forests forever? I just needed to figure out how to create my chronosequence on the outer coast.

Dying yellow-cedar tree.

CHAPTER 3

Fear and Forests in a Changing Climate

UNACCUSTOMED TO THE sweltering heat, I dripped in sweat as I darted around people on the Berkeley city streets a few days after I left Greg by the woodshed in Gustavus. Honking horns, busking musicians, the rumbling subway—the sounds of city life startled me after a summer of wild silence. I stared at the traffic, thinking about how emissions in one place affect so many others. I felt relieved to be on foot, and guilty for my flight south. I clenched a crinkled map of the University of California–Berkeley campus in one hand, trying to hide the evidence that I didn't know where I was going. Inside Mulford Hall, samples of tree species lined the hallway. The building smelled like eucalyptus, and a massive cross section of a redwood tree filled the corner of the foyer.

Sequoia sempervirens, I thought. *Same family—Cupressaceae. A relative of yellow-cedar.*

I ran my fingers across its surface—oiled and polished by the touch of many hands over time. Then I leaned closer, placed my thumb near the center, and began counting the number of rings from the tip of my nail toward my knuckle. I stopped short of forty, then stretched my arms out wide. My wingspan—just shy of six feet—spread across its diameter.

We lost these giants to loggers. I pulled at my shirt to stop it from sticking in the penetrating heat. *Could climate change take tree species as well?*

There is a climate threshold for every species—some point at which the conditions are no longer suitable for them. After learning about the yellow-cedar, I'd started looking at other species around me and wondering what their tipping points would be, and whether we'd let ourselves reach our own.

In Room 132, a girl with sapphire eyes, long blond hair, left uncombed, and a silver nose ring took the seat beside me. Her keys were hooked to her belt loop; they rattled against the chair. As she flipped through her notebook, I saw flashes of line drawings. She uncapped a felt-tip pen and began shading the triangle of a mosaic. Other students began to fill the seats surrounding us.

"I thought maybe I had the time wrong," I said.

"Berkeley ten!" she exclaimed, looking up.

"Berkeley ten?"

"Yeah, like the classes don't *actually really* start until ten minutes late. So come ten minutes late, you know?" She returned to shading.

"Umm, okay, that's new," I replied. "I'm only visiting for the quarter."

"From where?" She tapped the side of her pen with her forefinger, launching it in a complete circle around her thumb before it landed back in position.

"Stanford."

"Ohhhhhh, that makes sense," she said. "Hmm."

I shifted in my seat. "Yeah, over there they schedule meetings at times like seven past the hour. If you're not ten minutes early for someone who might be ten minutes late, you'll probably miss your chance to meet for another three months," I said. "So I thought I'd come early today."

She laughed out loud and extended her ink-stained hand.

"Kate, I'm Kate Cahill."

"Maddog, actually," the bearded guy next to her reported. "We just call her Maddog." I thought about inquiring further on that and then passed.

A man in standard forester attire—plaid blue shirt and blue jeans—entered the classroom, and the rustling of papers and backpacks stopped. He was wearing thin, wire-rimmed glasses, and he smiled with pronounced dimples on his cheeks and chin.

"Welcome to Applied Forest Ecology," he said. On the blackboard, he wrote his email and office number beside his full name, Kevin O'Hara.

Kevin and I had corresponded through email while I was in Alaska. I'd done my homework before asking for a spot in his course. His expertise combined forest management and stand dynamics—the scientific term for the study of forest growth patterns and the processes that shape them, such as interactions between species, and responses to fires or other disturbances. Through his class, I hoped to build a solid understanding of how forests develop, and that eventually, I'd pick up enough on field methods to choose my own—whatever I needed for the chronosequence I was designing around the yellow-cedar death.

Kate handed me a syllabus from the circulating stack, and I scanned the list of topics searching for what I needed. *Forest regeneration*—check. *Ecology of regeneration*—check. *Seedling production*, *stand density*, and *mixed-species stands*—check, check, and check. *Even-aged stands*—not so much.

The end of class, unlike the beginning, occurred as scheduled, and Kevin stood by the door to catch each one of us on the way out.

He stopped me and offered to meet anytime to help with my study.

"Door's always open," he said.

In the hallway, I finally asked Kate, "So, what's Maddog all about?" She blushed.

"Well, Kate here is vice president of the Forestry Club," her friend replied. "Every year we do a Christmas-tree cut. It's supposed to be a friendly fundraiser, but Kate is one maaaaaaad dog. She ran that cut like a drill sergeant."

"More trees, more money," Kate fired back. "I like efficiency!"

"That makes two of us," I said. She'd mastered forestry camp, an intensive summer program in the Sierras, and later I'd learn that the Maddog reputation wasn't only about cutting trees—she worked fast and hard in the field and made impeccable measurements.

I still hadn't unpacked the boxes I'd reclaimed from storage, and I was at least six months away from a study design. No grants for research yet. No experience measuring trees in a forest. No inklings about all the equipment I'd need, the weight limits of floatplanes, which temperature sensors could endure the harsh weather, and how long their batteries would last, or how many different ways there are to measure light in the forest or the amount of canopy above.

But I already had my eye on her for a field assistant. I'd need a team of research technicians to work on the outer coast, and grit would be my number one qualifier.

I READ EVERYTHING Kevin assigned in our months together and much more I selected on my own. I reviewed studies that used historical datasets for assessing change over time, then looked for others that used the chronosequence. I needed to understand the critiques and limitations of the methodology before confirming that it was, scientifically, still the best choice. It was.

Once I'd unpacked, my novels and creative nonfiction books sat untouched on my bookshelf. Instead, I adopted a strict diet of journal articles—first, covering everything I could find about the Alexander Archipelago and its forests, then looking for studies on other tree species affected by climate change. Somewhere in the stack, around mid-autumn, I came across a study conducted by scientists who'd revisited the journals of Henry David Thoreau.

I knew Thoreau for *Walden, or Life in the Woods*—his poetic documentation of two years, two months, and two days by Walden Pond, just outside Concord, Massachusetts. Published in 1854, *Walden* was one of the first American classics I'd devoured in college. Despite its accolades, I thought it was a slow read, but Thoreau's quiet, compelling meditation on simple living and natural beauty nonetheless stuck with me. I didn't know it in my twenties, not until yellow-cedar sent me searching, but Thoreau left more behind than transcendental ideas about humanity's relationship to nature.

In 1856, he wrote in his journal that he'd started "observing when plants first blossomed and leafed," and continued doing so "early and late, far and near, several years in succession, running to different sides of the town and into the neighboring towns."[1] Almost 150 years later, a scientist named Dr. Richard Primack went looking for his own data in the area to compare to Thoreau's, and the resulting study proved Thoreau's notes on plants to be as powerful as his published prose. Instead of space-for-time—the chronosequence I was considering on the outer coast—Primack had used a time-for-time approach by returning to Thoreau's rural landscape.

Concord was still largely intact, but more than two degrees Celsius warmer, on average, than a century and a half earlier.[2] To replicate Thoreau's observations, Primack and a colleague, Dr. Abraham Miller-Rushing, compiled information on hundreds of species. Their findings unveiled a suite of consequences of climate change. They documented a close relationship between flowering times and increasing winter and spring temperatures. Some plants flowered earlier than in Thoreau's day. Populations of native orchids, roses, buttercups, and violets, however, weren't responding to the temperature changes by flowering earlier; instead, they declined in numbers. Other species had disappeared from the area.[3] In a *Smithsonian* article, reporter Michelle Nijhuis wrote, "Even in the mythic American landscape of Concord, global warming is disrupting the natural world." She quoted Dr. Miller-Rushing as saying, "Now that we know what's changing, what are we going to do about it, and what are species going to do on their own about it?"[4]

There is a threshold for every species. That was beginning to be a mantra in my head. In an already relatively rare type of forest, yellow-cedar is a relatively rare species, but I was discovering that climate change's effect on it was not unusual. I care about trees, other plant life, and fauna, too, but that realization was also making me think more about people—What about us, our threshold, our future?

WHILE I WAS busy with classes, Paul Hennon and a mapping specialist flew to the outer coast before fall darkness settled in, when the harsh combination of sleet, snow, and rain would render aerial surveys impossible. He confirmed a long stretch of forests where the dieback had been active over decades—"old mortality to the south, and gradation of very recent mortality to the north," he wrote in an email. He wasn't sure about the exact distance. I got a quick estimate from Google Earth. It was about sixty miles by the straight line, much farther if you tried to trace the meandering coast.

Not craaazy, I thought. *Probably doable by kayak.*

He added (still with little punctuation): "the point, i think, is that it probably represents a spreading pattern." He confirmed what appeared to be yellow-cedar trees farther north that had not yet been affected.

So that was it—I had my study area. Slocum Arm on Chichagof Island for the long-dead and recently dying trees, then up into Glacier Bay National Park and Preserve for the healthy stands.

Muybridge, back in 1878, had placed cameras at different positions and watched the horse gallop by to create a portrait of what happens over time. Cowles, near the turn of the twentieth century, had studied sand dunes carved by wind in various stages of development to do the same. I'd center my observations on the yellow-cedar and the surrounding plant community; those trees would be my entry point into what lives and dies, where and when, and maybe why.

ON OCTOBER 12, 2010, I called John Caouette to talk about the pattern of tree death on the coast and my plan, still forming, to get there. The phone rang. I left a message. I wanted to discuss locations—Slocum Arm, Chichagof, and the outer coast of Glacier Bay National Park and Preserve. We'd emailed about the transect I'd selected, and he'd said it was a "fine idea," but a "tough place to access."

John never returned my call. The next morning, my cell phone logged a series of Alaskan numbers in the early hours—calls from many of the people I'd met because of him and the list he'd given me at the Paradise Café, the members of this widening circle of people connected to the yellow-cedar. A few left voicemails: "Call me back." Most did not. They all had the same tragic news. John had died in an accident running on a road in Minnesota. I was shocked by the strange and sudden tragedy, and torn up for the loss in his family and community. In the outpouring of grief that seeped in through messages and calls, the only conclusion I could reach for myself was that he'd planted a seed with me before he died. I would water it. I would watch it grow. I would see this through.

"I'm using the trees to show what's happening with climate," I recalled John saying in the rain. By the day he died, I, too, believed that trees could reveal far more than weather stations could.

AUTUMN FELL TO California winter, which quickly opened to spring, but my days passed slowly in tedium and mental exhaustion. So much reading. So much math. So many languages, human and computer, to learn and translate. An ecologist and a statistician and a computer programmer all have different words to explain basically the same thing. I was becoming all of those in one. To the forest ecologist, tree height and diameter are characteristics; to the statistician, they are just numerical variables; but to programmers, they become unique identifiers—[t_height] or [t_diam]—that can be used, through coding, to recall hundreds or thousands of data points in a stand or forest for analysis. Take a function such as tree growth over time, run some more math (in another coded computer language), and then, eventually, there's a vision of what a forest might look like in ten, twenty, or thirty years. Enter climate change, another complex variable or set of variables or even scenarios, and what was already complicated gets only more so. I kept plodding through the work, confused and challenged and at last rewarded until the same frustrating cycle would hit again. But I was generally convinced that I was learning the tools I would need.

The months flew by faster than I would have liked, given how much I still had to do to put together a proper plan for my study. I returned to Stanford at the end of Berkeley's semester, found a new place to live, again, and watched the clock tick down to the one narrow weather window I'd get come summertime to find and measure the trees on the outer coast. Paul said even the month of June could deliver stormy swells and thick fog, limiting the days I could reach potential sites. I made my way through problem-sets in statistics by day and spent nights troubleshooting the computer programs for analyzing the data-sets I had yet to collect. Early mornings, I studied the characteristics of a forest—from the amount of light reaching the floor, to the fullness of each tree's foliage, to the structure of old and young. Friday nights and Saturdays I guarded desperately to refuel my soul. Jonathan had taken a job as a field tech in Santa Cruz, a small coastal town about an hour south of the mountains from Silicon Valley. We went out for music. I fled to the ocean, jumping in the frigid, salty water to bask in the tidal rhythm. Waves, I rode waves those days. It felt far more natural than my own rhythm, shaped by long hours and screens. I climbed hills, meandered through chaparral, followed trails through manzanitas, oaks, and redwoods. Then Sunday—my Monday—returned, and I read and wrote and problem-solved.

ONE SCIENTIFIC STUDY in particular changed the way I saw the warming world in the months between Greg's pointed question, about whether the yellow-cedar was my muse, and my return to the archipelago. Described as "the first global assessment of recent tree mortality attributed to drought and heat stress," Dr. Craig Allen's research, published that year, linked increasing tree deaths and climate change.[5] Allen and his colleagues had combed through databases for studies around the world that surfaced from various combinations of key words, including "tree," "forest," "mortality," "die-off," "decline," and "drought." They contacted regional forestry experts to pull examples from government documents and other sources outside the scientific literature. What they found was widespread tree mortality—patches, just like the yellow-cedar graveyards, but of other species—occurring on all six plant-covered continents, and increasing in recent years.

I stared at Allen's figures in my Stanford office. Paul's map of the outer coast with the transect of the study area outlined in yellow was pinned to a bulletin board, along with the photographs he'd sent me from the aerial survey. Allen's article showed satellite images of each continent with white numbers placed at locations identified for every dieback. Inset photographs revealed dead patches of trees with lines drawn to corresponding numbers: #8 in Africa—a stark silhouette of leafless quiver trees (*Aloe dichotoma*) in the Tirasberg Mountains of Namibia, which looked like jumbo black corals standing on a bare rock face; #10 in Asia—dead red foliage dwindling on the Chinese red pines (*Pinus tabulaeformis*) in Shanxi Province, China; #5 in Central and South America—the luminously white trunks of the dead Dombey's beech, or coigue (*Nothofagus dombeyi*), in northern Patagonia, Argentina. I read through the results, then flipped to the appendix to take in the species list: the Atlas cedar (*Cedrus atlantica*) in northern Algeria, die-off from 2000 to 2008; oaks (*Quercus* spp.) in France, die-off from 2003 to 2008; ponderosa pine (*Pinus ponderosa*) in the American Southwest, die-off from 2000 to 2004.

What I call "the global synthesis study" certainly intrigued me as a scientist. I was fascinated by the patterns and impressed by the rigor of Allen's systematic approach. But it struck me even more as a citizen of this warming world, further awakening me to the realities of climate change—far more effectively than media headlines and news articles could do.

The world I saw on the page appeared so imbalanced and out-of-whack that I suddenly felt off-kilter, too. A tinge of anxiety pulsed through my body instinctively and hit my heart. I felt concerned for more than the trees. Around each dot on the maps, I envisioned a spiderweb connecting tiny squares for houses, huts, towns, and cities.

How many people witness the deaths of these trees?

How are people's lives affected?

I wondered what those patches meant for people across the planet, even people who might never know a tree or a forest but still breathe the oxygen they produce and live in a climate regulated by the survivors. Seeing those dots distributed north to south, east to west, meant that whatever I found in the archipelago might scale up. Scale up scientifically. Scale up emotionally. Maybe even scale up practically in terms of what people could do. Or what I could do.

"So the yellow-cedar is your muse," Greg's question echoed in my head.

No, I insisted again, this time only to myself. *It's a window into our future.*

John was right. The trees are telling us about climate.

But if the patterns of tree death were repeating themselves across the planet—as Allen's figures showed they were—so, too, might the responses of people and other plants in the years to come. The global synthesis confirmed for me that yellow-cedar wasn't the exception. Yellow-cedar faced ephemeral snow. The species Allen had listed were losing water and contending with extreme heat. But the more I dug, the more evidence I found of the bleak outlook for other tree species—such as the coast redwood, which relies on fog, but even the fog could fade away.[6]

What about us? What about people? I worried. *Is our own tipping point unavoidable?*

I don't know if it was the link to Thoreau, the startling maps in the global synthesis, Allen's list of species diebacks, or all the statistics, but the dying trees in Alaska didn't seem so strange or novel to me anymore. Increasing tree deaths, earlier spring flowers, some species in decline—I was looking at the new normal. The more I read that winter, the more everything I learned confirmed my first intuition about the yellow-cedar—it's a window into our future.

Instead of freezing in any sense of helplessness, I wanted to go back to Alaska immediately. Allen's study threw another log on the fire for me. It made me want to listen more. Observe more. Do more. Hope, a friend once told me, is like an untethered prayer. It's what we turn to when nothing we are doing and nothing we are trying and nothing we are striving for is working out. I wasn't interested in passive longing for some other future condition. Edward Abbey once said that "sentiment without action is the ruin of the soul."[7] To me, his words called for something more empowering than hope—a belief, a faith, perhaps, that what we do matters. The global synthesis made me want to find a way forward through far more than hope. Forward through action. Science, I believed, was what I could do.

Doubt stirred around the time the California poppies bloomed into bright orange bursts breaking free from the dry hillsides. They reminded me, daily, of the limited time remaining before my weather window opened in Alaska. I wondered if the grant proposals I'd written would come through soon enough. If I'd actually find a way to collect the data I needed across a series of forested sites along the outer coast. If the odds were too low for navigable ocean swell and tolerable winds. And if the methods that were finally making sense to me between white walls would just blatantly fail in the pouring rain. I launched emails back and forth with Paul and other ecologists who'd worked in temperate forests, first about science, but then increasingly, as summer approached, about remote logistics.

The first major hurdle was the tree stands themselves. *What stands would I actually measure, and how would I select them?* I needed some sort of proxy for time, some way to determine when the trees died. Coring trees to age and date them—only to then estimate time-since-death—could take another PhD, so that wasn't an option. Fortunately, Paul and others had come up with a way to date visual classifications of dead trees that decay as time passes. A yellow-cedar that looked like a telephone pole would have died, on average, eighty years ago, but one with dead foliage still attached would have died only three to five years ago.[8] Paul had photographs to use as examples. Maddog later drew the five stages.

Maddog's illustration of the five snag classes, used to classify the dead, decaying trees into categories for approximate time-since-death.

We'd need to survey the forests by plane first, to estimate the time-since-death of different stands, then go in by boat to our basecamps. *Boat captain. I need a boat captain. Car battery.* I added those to more to-do lists. *Test how much computer time we get off one car battery.* We'd have to reach each one by kayak and foot. *Kayaks. Add kayaks to the gear buy.*

A retired ecologist with the Forest Service told me to be humble, and to know "it's real easy to bite the big one" out there. The pressure of getting the data I wanted versus what was absolutely necessary would challenge me, along with tides, bears, and weather. He knew a bear-hunting guide, Paul Johnson (another Paul, so let's call him PJ), who had a cabin seven or eight miles north of Slocum Arm. It required a long crossing, exposed to open water, so I couldn't rely on it for steady use. PJ said I could take a few days of shelter there if I needed and claimed I would be the second woman in history to spend several months in Slocum Arm. The first one, so he said, had been stranded there in the 1940s while her husband went searching for gold.[9] He also warned me about sunny days: "If you get a warm spell, expect whipping winds and a heavy front to follow."

From the stands affected by the tree death at various points in time, I'd need to determine particular areas for study. Forest ecologists call these areas *plots*, and there are all kinds of approaches—fixed-area plots of a certain dimension, variable plots with different dimensions, nested plots with one or more nestled inside to serve different purposes. I looked up every approach, studied the pros and cons, and made informed decisions weighing both. I would use nests—concentric circles, like the outermost ring of sapwood on a tree and the border of the heartwood farther inside. The smaller circle would be for measuring small trees, because there could be lots of them. The full circle would be for measuring larger trees.

Once I was at my sites, I would study not only the yellow-cedars and the conifers but also the other plants on the forest floor, to get a sense of what was surviving, regenerating, or perhaps taking over as the trees died off. Decades ago, a famous ecologist named Alwyn Gentry designed a quick method for assessing plant communities.[10]

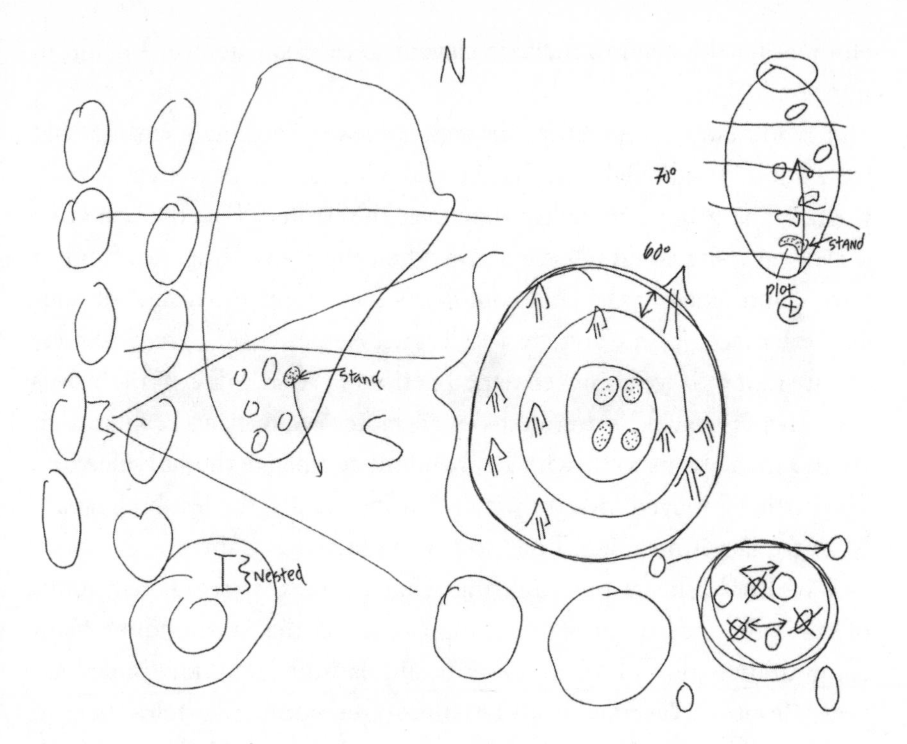

Early drawing of the plot design, made while I was considering various options. The multiple circles on the left indicate a chronosequence running roughly north to south; the nested circles on the right show ideas for measuring plants of various size classes.

Researchers since have used it all over the world. Pick a starting point, hold your hand out at a set height, and drop a plumb line so the weight reaches the ground. Identify every species touching the line that runs from your hand to the weight below and how many times it touches. Walk another meter. Do the drop again and again and again, collecting data across a fifty-meter transect. I could use Gentry's method, but other ecologists also use more time-intensive methods, such as estimating how much ground a species covers in a certain area, and doing that repeatedly for every species in many areas. Ecologists working in Southeast Alaska used the latter, so I decided I would, too. Science demanded so many decisions like this on a daily basis that fatigue set in. I created a standard rotation of "go-to" meals and daily attire, eliminating the need to make irrelevant decisions wherever I could in my personal life.

I read about temperature devices. I looked for small sensors that were durable and reliable so that I could affix them to dead trees or bury them in the ground to collect data over the winter. Together, the two sensors at a site could tell me if and when the snow came, how long it lasted, and whether the cold conditions had snapped beyond the species' tolerance. John Caouette's colleague, Ashley Steel, agreed to contribute money if we crunched data together. I ordered dozens of devices and ran tests in cold-water baths to check for accuracy and calibration. Ashley made plans to fly with Paul and me to finalize the site selection. The further I moved forward with scientific details, the less brain space I had for larger meaning. I had to get the methods right.

I built datasheets for collecting what I'd need to assess the status of the yellow-cedar and the composition of the surrounding plant community—plot number, description, latitude and longitude, aspect, elevation, distance from coastline, tree number, species, height, diameter at breast height (DBH, yes, that's a scientific term), dead, live, stressed. There were many more columns, each one of them carefully considered and justified regarding what they could tell me about the changes occurring. Every additional measurement would require more physical effort, and every effort required calories and time—all of which would be limited resources. I came across a figure from a study conducted in 1933 by an entomologist, F. P. Keen, who had been studying bark beetle attacks on ponderosa pine trees. Keen's diagram showed the branches and foliage of trees at various ages, some growing more vigorously than others. I added a column for crown fullness—a way to classify, visually, the health of all the green.

Grants trickled in, piecemeal. The first few gave me enough confidence to keep going, but not enough to know whether I'd cover it all. I was gambling with science on a shoestring. Worried I'd lose field technicians to other, better-funded projects, I hired mine before I had money to pay stipends or buy tickets. What I could pay, even if funding came through, was relatively minimal; the hardest work any one of us had ever done would be more like volunteer service.

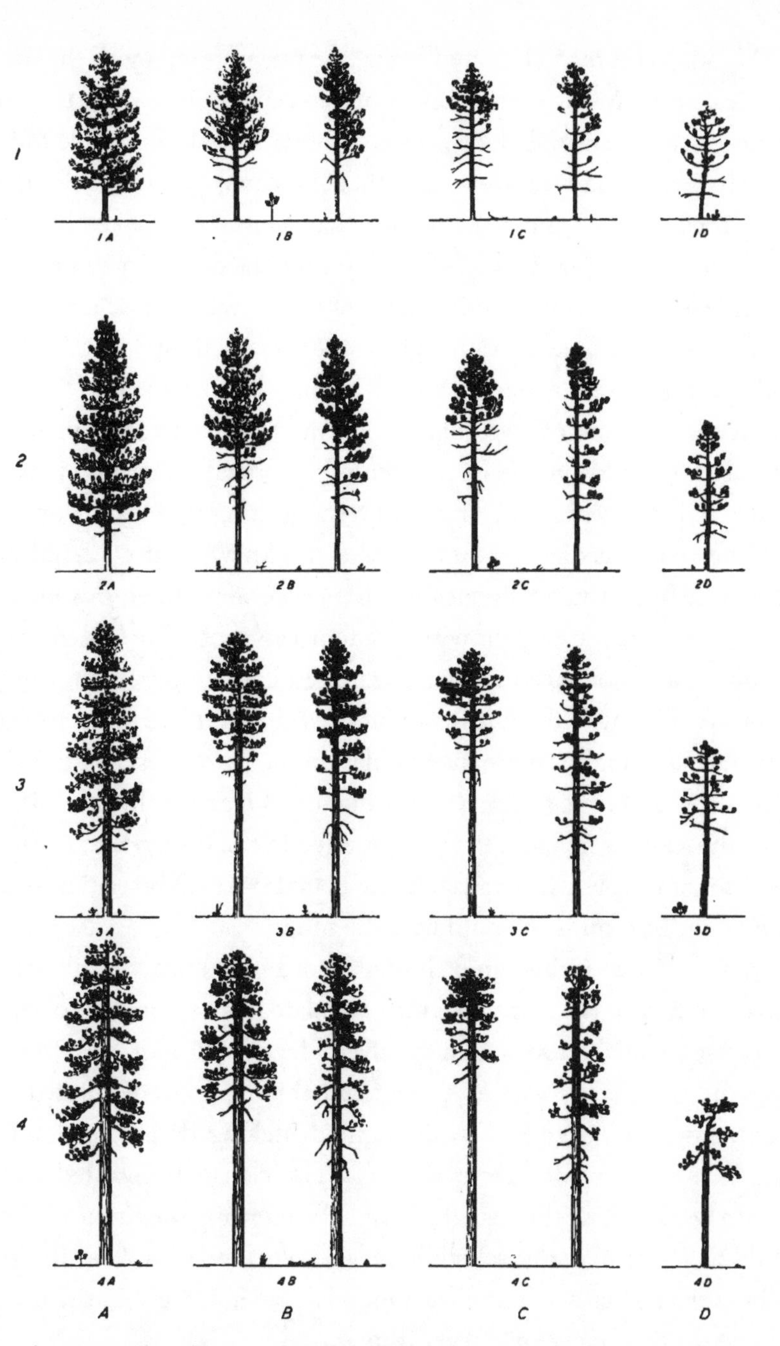

Keen's crown classification diagram. Reprinted from F. P. Keen, "Relative Susceptibility of Ponderosa Pines to Bark-Beetle Attack," *Journal of Forestry* 34, no. 10 (1936): 919–927, with permission from the Society of American Foresters.

Maddog committed immediately. She was unfazed by all the uncertainties, and said she'd never encountered bears in the wild, or worked in steady rain, or paddled, but she was pretty sure she'd be able to handle whatever situations arose, and that she could nail tree heights like a machine. I believed her. Sometime that semester I noticed black letters that read F-E-R-O-C-I-T-A-S on the inside of her right middle finger—"savagery" tattooed in Latin. Surely, it wasn't false advertising.

Tent space would be tight, given the weight limits on the planes, and the hull sizes in our kayaks if we had to move camp. I figured we needed a crew of two males and two females (myself included), because we could only haul two tents for separate quarters. I needed someone local, well aware of what he was getting into—someone who saw the yellow-cedar as a familiar tree and knew what cold and rainy in Southeast Alaska really meant. That was Odin Miller, or Óðinn, as he signed his name, a giant with size fourteen shoes. He had grown up in Juneau and now lived in a small cabin in Fairbanks without running water. A student at the University of Alaska, Odin had spent time as a field tech looking for methane bubbles on the frozen lakes of Siberia at fifty below zero. He spoke Russian fluently. Occasionally, he stuttered when he spoke English. He was interested in the history and culture of the Tlingit people and loved studying plant books. Odin said he'd carry one of our bear guns. I would take the other.

Kate I knew. Odin came through word of mouth via the yellow-cedar circle. For the last assistant, I circulated a post among colleagues and professional groups and received dozens of inquiries from all over the country, as well as a few from abroad. I weeded out the romantics whose cover letters highlighted lifelong dreams of going to Alaska, tossed out other experienced technicians who wanted a nine-to-five workday, weekends off, or time for morning meditation. I really couldn't guarantee any schedule at all. Weather would be the boss. Whoever said yes needed to trust me as a leader, believe in the project, and be ready to tackle whatever it might take.

Then Paul Fischer appeared in my inbox (of course, another Paul). He was pursuing a master's degree in forestry at the University of Washington, but he had experience inspecting large-scale energy projects, such as high-voltage power lines and natural gas pipelines. He'd assisted in surveys for rare species, and delineated wetlands in other projects, and he sold himself on the added bonus of strong skills in field mapping and GPS. He said he was burly—which a Skype interview confirmed, dark brown beard and all. Amidst asking me good questions about the science, the study design, and my strategy for managing a crew on the coast, he told me about biking cross-country alone, and scarring his shoulder with a hot wire that he'd clipped from a fence—and heated—to commemorate his success. When I told his reference, another scientist, where we were going and what we'd be doing, she said, "Hire him. He's your guy." I did, immediately. With his permission, I nicknamed him "P-Fisch" to avoid any Paul confusion with the one and only Dr. Paul Hennon.

I had my team.

Neither Kevin nor Paul Hennon would have signed up for a full summer on the outer coast, but a short trip was enticing and manageable around their work responsibilities. They both volunteered for a week. By then I'd learned that Kevin, knowing how species interact with each other under various conditions, could look at the undergrowth of a forest and visualize how it would grow and take shape over time. His years of studying forests across the globe gave him a comprehensive toolkit of the field methods, so I knew he could help me figure out what would work best out there. Paul, with his regional expertise, could identify absolutely anything green, and he knew how to work in the tough, temperate conditions we'd face. I jumped at the opportunity to have these two exceptional scientists with my team at the start of our journey. The first time we'd convene, all six of us—Paul, Kevin, P-Fisch, Odin, Kate, and me—would be on the outer coast.

The last chunk of funding came through from the National Park Service one week before my flight north in mid-June, ten days before my flight with Paul and Ashley. I would have left much earlier, but classes end late at Stanford, and that added another layer of time stress to my already pressured situation. I went to campus to pick up the check, which came via FedEx with less than twenty-four hours to go. I was standing at the copy machine in the office at Stanford, printing an extra set of datasheets on Rite in the Rain paper. Helen, one of the program directors, found me there. She had watched the months of my determined preparation, processed receipts for unusual gear purchases, and signed off, like my advisers, on a plan for a PhD that no one called "risky" out loud but everyone knew was.

"Are you coming back tomorrow before you leave?" she asked, speaking quickly and quietly by my side. Other students and staff filtered in and out of the office.

"Not sure yet. I'll go over my lists again tonight, but I think I have everything." I had already shipped over 200 pounds of gear and food by boat and air cargo for the months ahead. It was cheaper than paying for extra baggage on flights. I'd check another 150 on the three-bag limit.

"You should get out of here today," she said. "It seems there's a bit of a fire starting. Lots of questions circulating about safety. Probably triggered by whatever you collected around campus. I'm not sure. I thought I'd deflect concern rather than inquire further. Seems like you've got this."

"Totally," I said, terrified that last-minute bureaucracy would try to stop me.

Months prior, I'd gone hunting for a gear room or any list of fallow equipment on campus. Both existed, and they were carefully maintained by a health and safety coordinator. I'd tracked him down, and he had quizzed me on my plan. I had backcountry medical training, I assured him. At least one other team member did as well. We'd carry Forest Service radios to report our location on a daily basis. If we missed two check-ins, they'd send a search team. (It could still take days for anyone to reach us, but I didn't tell him that part.) We were lined up to meet with a bear biologist in Sitka and a fly-fisherman in Juneau; the latter had been charged by bears more times than he could count. We'd implement their recommended protocols for avoiding attacks. Odin had already completed firearm training, and I was next. The safety officer at the Forest Service had asked me about approvals from my "unit plan's safety officer." *Unit plan? Safety officer?* I think this guy at Stanford was technically that. But as far as I could tell, I was the first student he'd come across who was planning to carry out remote research by kayak, floatplane, and foot in bear country. There wasn't exactly a unit plan—something apparently standard for the Forest Service crews. He helped comb the campus for me. I walked away with a satellite phone, a waterproof box for radios, GPS units, multiple first aid kits for minor cuts and bruises, and then a major one for absolute crisis.

"Let's just say it might be better if you're already gone tomorrow," Helen said, almost whispering.

"Got it."

She had that look in her eye that I knew only from my mother—the soft, supportive gaze of the unsaid be safe, make good choices, go, go now.

I collected the datasheets from the printer. I held a big stack with thousands of blank cells awaiting numbers from forests I had yet to see. Jonathan took me to the airport and kissed me goodbye. And then I was gone.

CHAPTER 4

Solving Puzzles

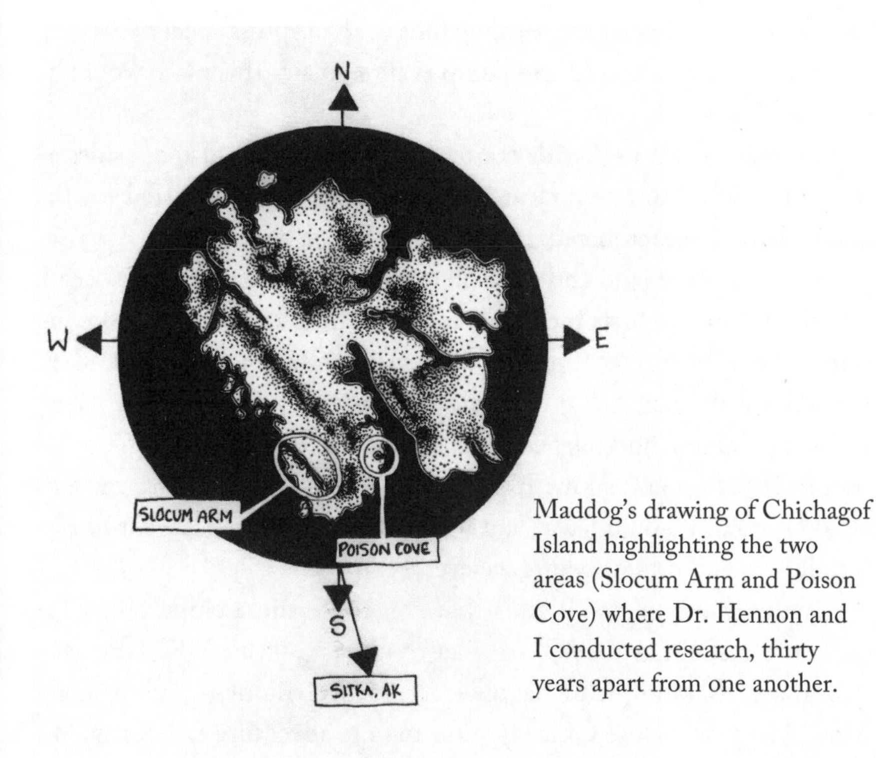

Maddog's drawing of Chichagof Island highlighting the two areas (Slocum Arm and Poison Cove) where Dr. Hennon and I conducted research, thirty years apart from one another.

IN THE TOPOGRAPHIC maps and marine charts I'd stared at for months, the Khaz Peninsula looks like a finger jutting out from the base of Chichagof Island, a long sliver of land with a series of mounds cutting between the Pacific Ocean and Slocum Arm. On paper, the peninsula appears as just another feature among the thousands that form the winding coastline. But from the ocean view, the head of the Khaz—the rock face that drops into turquoise water at the finger's tip—stands prominently on the outer coast like the Gates to Valhalla.

Rounding the Khaz by boat is a delicate, decisive dance between peace and peril. The surface of the rolling sea shatters against sheer cliff. Swells surge between rock islands. The ocean twists and twirls, shaped in part by the complicated contours hidden below. In the forever moments between open and protected waters, maelstrom and reprieve, silence is instinctual. So is the exhale of relief, as troughs and peaks drop away to glass. Then there's forest, light green and dark green and darker green, and dead trees—skeletons of the cypress species.

I inhaled the salty cold and damp evergreen air. This place would be our summer home.

On June 26, 2011—five days after that flight with Paul and Ashley—the clouds lifted to reveal clear blue sky on the day I departed for the outer coast. I remembered PJ's warning about sunny days. Air rises from the warming land and sucks in the cooler, wet air from the ocean. Winds and steady gusts inevitably follow in the stormy front. Looking across the open waters from the dock in Sitka, one island to another, I could see the sunlit top of Mount Edgecumbe, the volcano where Dr. Paul Hennon and his team had mapped yellow-cedar trees in relation to elevation and snow. If I had known how the next two months would unfold, I would have slept for a week and eaten pasta for breakfast, lunch, and dinner before we ever left town.

Of the thousands of islands that constitute the archipelago, Chichagof stands out as the "C" in what residents call the ABC Islands—Admiralty, Baranof, Chichagof—revered for wildness and wildlife. Most Alaskans choose Chichagof for remote adventure or bounty, like fish, berries, and deer. Locals in Sitka had told me the bears on the outer coast were some of the biggest in the archipelago. I should expect the unexpected; they seldom interacted with people, so there was no way of knowing how any encounter would go. It was not for Chichagof's alluring remoteness that I picked the wild, west side of the island. I needed it because we already knew it had patches of standing dead trees as well as other areas with signs of more recent stress. But because the aerial plan had failed for selecting sites, I'd have to create the chronosequence by boat and foot.

"Yesssss, push off. I love push-off!" I announced, anticipating the moment when we'd finally leave. We were still lugging hundreds of pounds of gear down to the boat I'd chartered. Kate, Odin, P-Fisch, and I were going in first to set up camp. I needed the weather to hold because Kevin and Paul were scheduled to join us the following morning in a second load.

Time with Paul and Kevin would be limited and focused on resolving any last issues. They'd fly out after five days, leaving me and my crew for at least another eight before we headed back to town for the first dose of our brief "dry-outs." I still had dozens of scientific decisions to make before I could be in go-go-go mode with my crew for the months ahead. Once Paul and Kevin were gone, there'd be no more asking questions. We didn't have Internet or cell service. There wasn't space for heavy reference books. Satellite phone coverage was spotty. My radio only connected to the Forest Service dispatch in Sitka to confirm our daily whereabouts.

Hauling bags down the dock, I tried to focus on being present and not worrying about what the sun would bring next, if Kevin and Paul would make it out tomorrow, and whether we'd fill all the empty spaces in my datasheets. I was eager to untie and go, finally go. I'd been going for months now without ever actually getting there, and now—finally—the trees were just a three-hour boat ride away.

Kate smiled and directed Odin to grab the other side of a metal "bear-proof" box packed with dried food.

"What we have we have, and what we don't we don't," I said. "From here on out, we make it work."

The new plan, written in pencil, was as follows: We'd get out to Slocum by motorboat in two waves—one for me and my field crew and one for Kevin and Paul and the kayaks. I still needed to choose the sites we'd measure, and doing so in a random but systematic way remained a fundamental prerequisite for ensuring a valid scientific study. After the dispiriting flight over Chichagof, I'd come up with a new plan: The boat captain scheduled to transport Kevin and Paul in round two would stick around. We'd survey the forests along the coastline with a motorboat for a day or two, then select our sites from the shore.

I planned for my crew to stay out in the wilderness for twelve- to fourteen-day stints (a duration determined by another set of unpredictable variables), and we'd travel from our basecamps back to town by whatever transportation was possible that day—boat if the fog layer was too thick to fly, floatplane if the ocean swell was too big. We'd leave the kayaks out in Chichagof for the summer until we reached mission complete, or the weather window closed, whichever came first.

We had eight weeks to make four trips happen. I estimated ten to sixteen days per trip, leaving a couple days on each side of the target length for wiggle room. We'd need flexibility for weather that could strand us at basecamp, or, worse yet, in Sitka while trying to return to Slocum. Setting up and breaking down camp in new locations meant that two days per trip would be useless for data-gathering. I carved out time in our schedule for two- or three-day breaks in Sitka, but I was already thinking so few days in civilization would be tough. We'd repair gear and restock food, charge batteries, wash clothes, sleep warm, and eat as much as possible before doing it all again. I saw flying or boating back and forth as a waste of time, money, and fuel—fossil fuels—so we'd stay out as long as we could on the coast before supplies ran short.

PUSHING OFF FROM Sitka came with a surge of relief. For those few hours there was absolutely nothing I could do, no troubleshooting, no checking or rechecking, no loading or calculating. I could just be, quietly, in the space between.

I took the seat across from Charlie, our captain, and stared out at the blue horizon as we cruised northwest of town, passing a sprinkling of islands I'd studied on the charts. We entered a long string of narrows, constricted by rock and forest to the east and west, and then, like passing through a funnel, but reversed, we emerged into the big, wide-open Salisbury Sound—our first taste of the outer coast.

Winds swept out of the passes, colliding with seas. We followed the waves up and down, our hull smacking against each trough. Charlie tucked in behind a craggy, triangular island called Klokachev, a massive chunk of bedrock exploding upward a thousand feet from the sea. I knew exactly where we were from the maps I'd studied.

"Leo Anchorage," Charlie reported, confirming my thoughts. From the office, I'd appreciated that spot on the map for its name, which serendipitously bore my initials. From the water, I relished it for the moment's reprieve it offered along the lee shore. It was the last protection before committing to the Khaz.

Charlie pulled back on the throttle, slowing down to regroup.

"Ten miles until we reach protected waters," I said.

"You still want to go?" he asked, raising his furry white eyebrows. "Because between here and Slocum there's nowhere to stop."

"I still want to go."

Charlie pushed the throttle forward, launching us into the rolling sea again. An otter popped up, clutching kelp, and then dropped down, leaving behind a ring that dissipated on the surface. At the top of every wave, I saw the Khaz getting closer; in the troughs, only sea. We rode in silence until we were out of the open and onto glass in Piehle Passage, weaving through rocks and swirling currents to round the Khaz. And then we were there, finally there inside Slocum, cruising by yellow-cedar, spruce, and hemlock, searching for a spot to call home, temporary home, Basecamp #1.

The consecutive days of sun and manageable swell that allowed Paul and Kevin to arrive felt lucky—the kind of luck I needed. I savored the fleeting blue, knowing that white fog, wind, and rain were next. Kate and Odin stayed back in camp to practice plant identifications. That left me and "the boys," as Kate pointed out, for the boat survey—the first step toward creating the chronosequence. We would survey the trees from our near-shore perspective and look for stands of yellow-cedar farther inland on the hillsides. Each observation would be marked on our GPS with a code indicating time-since-death. Then, across the chronosequence of mortality, we could randomly select sites at various distances inland from the GPS-tagged coastline. If Plan B didn't work for selecting forests with yellow-cedar trees that died at different points in time, I didn't know what would.

"Just tell me what you need," Scott, our second captain, said, as I stepped aboard. "I'm the boat captain, but you're leading the science. How fast should we go? How far do you want me from the shore? You make the calls."

"Great," I confirmed. "Let's get P-Fisch on one GPS unit, and I'll take the other. Paul, seems like you and I should be on observations. We both make each one silently, independently, then report back, reconcile any differences, record it, and move on." Kevin volunteered to take notes on paper for backup. I held my binocs in one hand and the GPS in the other. I rested my thumb lightly on the waypoint button so I wouldn't have to look down to mark locations.

"I'm not going to be able to stop," Scott said.

"Really?" That was new information for me. He nodded.

"Okay," I replied, frustrated. "Then we'll need to figure out a speed that's slow enough for us to record but not so slow that we get pushed around by the swell, right?"

"Exactly. Five knots? Slower?"

"Let's test it," I said. He tried a few different speeds until I picked one.

"You're on trees," P-Fisch said. "I've got distance covered. I'll call it every hundred meters. We both mark waypoints." He drew lines down my waterproof notebook for data columns and then tucked the pencil behind his ear.

"I'll record locations in the GPS unit but forest status on paper," he reported.

Then we motored south to north for hours along the coastline, staring at trees through binoculars to discern species and identify branch decay on a boat swaying in the sea. Paul and I called out stages of tree death.

"Mark waypoint," P-Fisch commanded.

The view from the boat, showing class four and five snags indicative of old mortality. Photograph by the author.

"Telephone poles. Hardly any branches. Call it O. Old mortality," I reported.

"Mark waypoint."

"Again, O." Paul chimed in.

"Mark waypoint."

"Looks like we've still got main branches. Call it M. Midrange."

"Mark waypoint."

"Recent, very recent death. Still dead leaves on all the branches. What do you think, Paul?"

"Agree. Call it R."

We mapped dead trees for miles and miles. I kept tunnel vision on the job. Those markers had to be accurate.

Collecting data points took a day and a half. Scott dropped us onshore and left immediately, with dark skies impending. P-Fisch and I spent hours in the tent entering data into my field computer, another score from the gear salvage at Stanford. Once we had all the information from our datasheets in the GPS units, from there on out, it would be a treasure hunt—a geocache. We'd have randomly generated locations—potential plots—stratified into forests affected at different points in time. When we finally managed to load the units, with only minutes remaining on the car battery powering the computer, P-Fisch and I cheered so loudly that our voices echoed across the cove. In those moments of challenge and fleeting victory, it was easy to let the demands of science distract me from the bigger picture—climate change as the increasing threat to us all, and what those trees might mean for humanity.

We wasted no time. The next morning, I stood on the beach with my crew and Kevin facing the inevitable jump through "the wall," the vertical line of brush where dense forest meets rocky coast. The temperature hovered in the 40s. Cold wind swept across the inlet, driving rain sideways and sea upon the shore. I pulled my wool cap over my ears. Kate adjusted the suspenders on her army green rain pants. Odin took one of his giant, 6′4″ steps over a washed-up log and scanned the wall from top to bottom looking for an entry point.

"Where'd Paul go?" Kate asked, tucking her braid inside her jacket collar.

"In, he's already in," I replied, tripping over a rock. My thick rubber boots kept water out, but I was clumsy with their weight.

"That man moves like a forest fairy. He makes it look so easy," I said, annoyed with my footing. I heard the foliage rustling behind the wall where he'd disappeared, then silence.

"So, if you're estimating the percent of stressed or dead foliage," Kevin said, pointing to a tree with branches of brown foliage, and others with a mix of yellow and green, "are you considering the yellow, too?" He was patient and quiet, but when he spoke, he shot sharp questions I knew were critical to resolve.

Brown and yellow? Brown only? I thought. *Brown is definitely dead. Yellow is likely stressed but might recover. That's more of a yellowy-green than a yellowy-brown.* I squinted. *Is that orange? Is there a right answer? I don't know. Does he know?*

"Yellow, orange, brown—the rust tones. All classifiable as stressed or dead," I announced. Kevin nodded and took another step on the beach toward the green wall. I double-checked the safety on the pistol strapped across my chest and slid the bear spray on my hip belt around to the front. Odin secured the .338 bolt-action rifle over his shoulder.

"If you shoot, you shoot to kill," an experienced hunter had told us in Juneau. "Only injuring a bear is not an option. If you ever decide to open fire, you have to take it down before it takes you down."

I was far more worried about science than bears. The moment we finished this first plot together, there'd be no turning back. At that point, the methods used at every subsequent plot had to match the first one exactly. Backcountry logistics would take a backseat to science, and science demanded repetition, accuracy, and consistency. After Paul and Kevin left, the game for Kate, P-Fisch, Odin, and me would be twofold: survive (and try not to bleed out or break any bones), and work like machines for the rest of the summer.

"I've got the clipboards, the datasheets, and the Impulse," Kate announced. Relying on basic trigonometry, the Impulse allowed us to measure tree heights by laser. Unfortunately, the process was not the simple point and shoot I thought it would be: you had to shuffle around the forest floor until you found a spot with a line of sight to both the base of a tree and its top; shoot the laser to the tree bole and press a button to measure the horizontal distance; shoot to the treetop and press again; then shoot to the tree base and press one last time. The Impulse calculated the distance of the missing leg: the tree height. Finding the sightlines to the base and the top of each tree took a lot of traipsing around. I had backups for most of the equipment if it broke, but another Impulse had been too expensive for my shoestring budget. If the Impulse failed, the whole project would fail.

P-Fisch looked down at the tiny screen on the GPS unit, rubbed the surface of its plastic protector to wipe away the fog, and pressed a few more buttons.

"I marked our starting point," he said. "I've got a distance to the first plot. We're good to go."

Kate went first with a duck-and-enter strategy. Odin extended his arms out in front of his face and charged straight in like a bull. I curled my head down, protecting my eyes, and pushed forward as I spread apart two alder branches with my bare hands.

Gardening gloves, I thought. *Add those to the list of supplies for the restock.*

Once inside the forest, we picked up the pace together, chasing after Paul to catch the grace that came from his decades of studying these forests.

Between the rain, the water on the plants, and my own sweat trapped inside my supposedly breathable Gore-Tex jacket, I was completely soaked by the time we reached the location for our first potential research plot. Standing in the middle of the forest, I practiced the protocols for assessing whether the surrounding conditions met all the qualifications for including it as a study site. If not, we'd reject it and hike another random distance to try again. Safety-wise, we were good to go; it was steep (far steeper than I would have liked for our first plot), but by the reading on my clinometer, the slope was not steep enough to abandon it altogether. Holding a special glass piece to my eye, I rotated slowly in a circle, stopping at every tree that came into view through the prism. Based on the extent to which each trunk appeared displaced in the glass, I could tell if the tree was too small to include in our count. Kate kept tally beside me, noting the species of each tree included and whether it was dead or alive. After I completed my 360° panorama through the prism, Kate added up all the "in" trees. The process enabled me to quickly assess basal area—the space covered by tree trunks. If the basal area was too low, we'd deem the surrounding landscape unsuitable as a forest for my study and move on. Kate checked to see if the majority of trees were yellow-cedar.

"We've got a keeper!" she said. I was cold from standing still and relieved to hear we could begin working to get warm. Kate grabbed a stick, tied a strip of pink ribbon called flagging around one end, and shoved the other end into the moss at my feet.

"Plot center," she proclaimed. "Number one of forty. Here we go!"

Penned in by too many shrubs, we started off like a herd of turtles, slowly bumbling around one another with none of us knowing which direction to go. Odin fumbled with various tape measures in his pack. P-Fisch and I futzed with our GPS units to record the plot coordinates. Kate grabbed the laser Impulse and began shooting the same tree repeatedly to check accuracy.

Kevin and Paul stood like coaches on the sidelines, observing everything but refraining from giving too much direction. My stomach grumbled. I checked my watch. It was almost time for lunch, and we hadn't actually recorded anything yet.

Kevin tapped my shoulder from behind and said, "Sloooowww is to be expected." I wiped the raindrops off our datasheets. The paper felt saturated.

"The strategy is to get it right this first time," he added. "Then you've got a whole summer to get faster."

Forget Rite in the Rain paper, we need whatever oceanographers use underwater, I thought, making another mental note for the restock.

"It'll work," Paul said, noticing me testing pencil lines on the soggy pages. "Press hard but not so hard you tear it. Dry them out at night. Photograph them in the tent so you have a backup."

Wipe down Impulse. Oil gun. Digitize datasheets. I took a deep breath and created a new mental checklist for nightly camp tasks. *I do love lists!*

To measure plot diameter, Odin tied the end of our transect tape to the center stake and then pushed through blueberry brush with the tape trailing awkwardly behind. Once we had the pink flagging secured in a large circle, which bounded our forest to the fixed area, we could finally begin identifying trees and sizing them up. Odin's world would be confined to a one-by-one-meter square delineated by white PVC piping on the forest floor. He'd identify every plant species it contained, measure their average and maximum heights and the amount of ground each species covered, and then move the square to a new location until he had eight quadrats completed per plot.

Paul volunteered to help with tree diameters, which I'd later separate into size-classes ranging from saplings to big trees. Compared across the chronosequence, the measurements could show me which species grew up as the cedars died off. It took three sets of eyes to ensure we didn't miss a tree. We spread out in a line like the big hand of a clock and slid our way around the dial, tree by tree. P-Fisch led, nailing a numbered aluminum tag into each one. Paul measured each diameter with a cloth tape. The bigger trees required two people—one person to hold the first end while the other walked the tape around the tree. I hollered "Spruce" or "Western hemlock" identifications. Paul jumped quickly for the cedars. He'd lay his hand on the rippled bark and exclaim: "We've got one," and all of us knew what "one" meant. He'd step out from the trunk for a moment and gaze up, his bright blue eyes filled with admiration, regardless of whether the tree was dead or alive.

There, underneath the cedar carcasses, I wondered if loving this species was what had carried him through the thousands of measurements he and his colleague had made to solve the mystery of tree death in the archipelago. Paul's puzzle had been figuring out what was killing the yellow-cedar trees. Mine would be figuring out what happened next, not just for plants but for people, too.

Ten miles across the island from our first site in Slocum Arm—and thirty years earlier—Paul's career-long endeavor had begun in Poison Cove. He had chosen the cove for its edges—the abundance of places where patches of dead yellow-cedar trees abutted groves of live ones—so that he could move efficiently between the two to study them both. Like an infectious disease doctor at the site of an epidemic, Paul collected as many samples and recorded as many observations as possible. His mission was to document where on the landscape the cedars were dead and dying and where they weren't, and to describe the signs and symptoms of the ill.

Paul and his wife, Susan, spent two summers at Poison Cove down on their hands and knees in the mud, digging up the roots of dying trees—and healthy ones, too—for comparison. The published papers I had read reduced months of work to essential details: "The roots . . . were excavated to study symptoms and organisms associated with decline and death." And, "Of the 1,864 isolations, 1,047 (56%) yielded fungi."[1] Only

Paul and Susan would ever really know what "excavating" entailed—that their hands would be covered in no-see-um bites and their days confined to the radius of a tree's roots. One thousand, eight hundred, and sixty-four "isolations" meant slicing tiny pieces from roots they'd ripped out of the ground, then, using petri dishes, growing whatever fungi might be present. They worked in a remote hunting cabin (not exactly your standard lab setting). After seasons of dogged investigation, the usual culprits in any other mass tree death were found to be innocent. Fungi, viruses, beetles, and molds—none of them was the killer.

A mystery that Paul thought would take his PhD to solve took his career instead. It was only out on Chichagof that I could really understand what the thousands of observations he had made over all those years required—how much of his life and energy he'd given to this painstaking, incremental, yet electrifying work. No one suspected climate change as the culprit early on, and no one would have accepted it without ruling out everything else along the way. Would I, too, work a lifetime to find an answer? Would the answer to what happens next be enough for me?

On our final afternoon together, Paul and Kevin stopped asking me questions and became extra workers on the team instead.

"Write it all down now," Kevin reminded me. "You think you'll remember every decision, but you won't. If you're only including saplings taller than one meter in your counts, note one meter. If you're shooting photographs of the canopy at an aperture of F11, note F11. Your methods make your bible."

We measured tree heights and diameters. We estimated the percentage of dead and stressed foliage. We noted where each stood in the canopy in relation to its neighbors. The cells in those empty datasheets slowly began to fill, but the big stack I carried indicated all that was to come. Paul fell into a groove of less teaching, more doing, and I felt my confidence build. They were letting me go, and I was ready.

At night, we ate dinner standing on the beach in the evening drizzle.

"No big tarp?" Paul asked, gazing around our basecamp cove for shelter from the rain. We cooked on the beach in the tide zone so any remnants left behind would go out with the sea instead of lingering for the bears. In our minimalist, leave-no-trace approach, we used small backpacking tents, and we tucked them into the forest across the cove.

"No big tarp," I confirmed.

There wasn't time for sitting around, so I had thought a big tarp would be excessive. "We've got a small one we'll put up in the trees over the boxes. That way we can at least go food shopping under cover."

"Hmmm," Paul said, holding back in front of my crew.

"What did you do for cover at Poison Cove?" I asked, searching for any last tricks I hadn't considered.

Paul paused. "Well, we didn't do it like this."

In the minutes since serving, everyone had devoured the rehydrated curry. The outer coast had redefined hunger already. What I used to know only as a faint reminder from my stomach became an aching pain throughout my body—belly and muscles screaming for fuel and seldom satiated.

"What do you mean?" I asked, savoring my last bite.

"We didn't have to. We had it cushy with a cabin and a stove. We based there all summer." No one said anything. I imagined clothes lines strewn across log rafters for drying gear and rubber boots warming by a fire—so lavish. I'd already developed a meticulous system for tent entry, exit, and clothing maintenance. At night, I'd stand outside the tent, rapidly remove my rain pants and jacket, and crawl into the front vestibule wearing only long underwear. My wool was soaked, so I'd pull the tights down around my thighs and back into the tent—naked butt first. I'd sit at one end and strip wet layers to avoid getting the floor or my sleeping bag wet. Then, finally, I'd switch to my wool top and bottoms that never left the tent. Just knowing I had one set of dry sleep clothes enabled me to tolerate the daily cold and wet discomfort. P-Fisch wore his wet clothes at night to dry them out with his body heat, but I couldn't generate enough heat to dry anything. Putting the cold, moldy wet ones on in the morning was the worst. Odin claimed his toughest challenge waking up was smelling his socks and picking the two least disgusting. He hung them in the tent, a sure bear deterrent.

Odin leaned over to see if any food remained in the cooking pot. It was empty.

"Well, we have a nest, and pretty soon I'm going to crawl into it!" Kate said. She headed toward the salt water to dunk her bowl and never called our one dry spot a "tent" again.

In the morning, I woke to the clanging of someone packing tent poles. Outside the nest, I watched Paul and Kevin move their gear down to the beach as I pulled my rubber pants over my boots. I was drained and looking forward to leading my crew without being under observation, but I also knew a wealth of knowledge was leaving with them. I felt like a kid about to be left home alone for the first time—pretty sure I'd be fine, ready to celebrate, but uncertain about the problems that could arise.

We heard the plane long before it landed on floats outside our camp. Kevin and Paul were packed and waiting on the rocky beach. The pilot stepped out of the plane in waders, and we watched as he held on to a float with one hand and shuffled gear with the other. Bush pilots on the outer coast rarely let a passenger load a bag, but this isn't chivalry. Knowing how weight will influence flight, the best of the best hold each item that comes onboard, make an estimate, and choose exactly where in the aircraft it should go. The pilot pulled the plane in toward the beach to shallower water so Kevin and Paul could load up. Water was just about to seep over the top of Paul's rubber boot when he leapt in.

"Remember," Kevin said to me, "safety first. Eat. Stay warm. Be careful." He reached into his jacket pocket and handed me a few granola bars. He'd already passed on his oatmeal portion at breakfast and said he'd eat back in town. The plane took off, leaving the four of us to finish plot number one and forge on toward number forty. I ran back to the nest to grab my daypack. Inside, Kevin had left his blue fleece, folded neatly, on Kate's sleeping bag.

By the time I returned to the beach, Odin, P-Fisch, and Kate were packed and ready to go.

"Just us now," P-Fisch declared. "We've got a job to do!"

Kate let out a long howl. Queen's "We Will Rock You" popped into my head. Boom boom, clap. Boom boom, clap. I felt like a real scientist for the first time. Technically, it would be *my* dataset, but I felt that my assistants—my teammates—were just as committed. I wasn't the only one who wanted to know what happened next. We were alone on the outer coast with thousands of trees awaiting us.

It took us two and a half days—another full day after Paul and Kevin left—to get the first plot right. We needed to measure ten plots per trip. Two and a half days for one was way too slow. The only solution was what Kevin proposed—get stronger and faster.

Around nine hours into plot two, wet and exhausted, we debated whether to finish for the day, or bail and return in the morning.

"If we go out, we're better rested for tomorrow, but then we have this lingerer. We'll end up staggering plots over multiple days. Seems inefficient," Kate said. "I don't like lingerers. I like done."

P-Fisch tagged a yellow-cedar with three hammer strikes, and then said, "If we stay in, we save the calories we'd burn on the extra hike here and back." He didn't stop working for discussion. He never stopped working, really. He flung the measuring tape around the trunk and pulled it tight.

"Sleep, oh sleep sounds so good. But, well, we, umm, we don't have extra calories to waste," Odin debated. Given Kate's disdain for inefficiencies and P-Fisch's innate decisiveness, I was slightly concerned already that Odin's process of thoughtful deliberation would drive them both crazy. I had intentionally brought together a contrast of approaches and styles on the team, thinking we'd keep each other in check and avoid making any rash decisions that way.

"We'll have to be loud for the bears if we hike down at night," I said. I asked if anyone had any safety objections to either option. No one did. "Okay, let's avoid the lingerer," I decided.

We stayed and finished in twelve hours total, but that was progress—a day and a half less than before.

The sea turned to glass on our paddle home that evening. Kate had thought our blue double kayak was the color of a whale, so she had named it accordingly on day one. The boys took the fire red one, which they had named the Dragon. The rocky cliffs reflected on the water's surface, creating the illusion of another terrestrial world on the plane below. I felt dizzy and disoriented trying to discern which way was up.

"Whoooaaaa," Kate said up front, snapping out of her own tired daze. "Jellllllllyfish."

I squinted to adjust my focus beyond the reflections to the ocean beneath me. The squirmy body of a large jellyfish squished forward and then puffed up like a pink umbrella.

"Poooooooof," Kate said.

In some strange, twisted way, I think we started to like the challenge. Plot three: ten hours from the jump into the wall of forest and out. Odin began clocking his rate on understory quadrats. We plateaued at eight hours for plots four and five, and then plot six crushed us. The forest was thick with small trees about our height, and the hike in was our slowest yet. We pushed through saplings, unable to watch our footing. I stepped on a mound of moss beside an old, downed tree and felt the ground give way. I could hear the earth tearing as I fell through. I landed on a heap of roots inside a massive hole created by the tree that had toppled over.

"Whoa!" Kate exclaimed. "Are you okay?" She peered down at me from above.

"I think so," I said, wiggling my legs to check for pain and looking down at my body for blood. One of the roots had ripped a hole through my rubber pants, but it had left my skin unscathed.

"That's a good way to break a leg or slice an artery," I said.

I'm so glad Mom can't see this.

"So, big downed tree means big hole somewhere, even if we can't see it," I added.

Kate laughed. "Noted." She pulled me up to moss-level from underground, and we continued on.

Our first random location landed us in a boggy spot without enough trees to measure, so we rejected it and hiked on. The second attempt landed in a steep ravine. Strong winds must have hit because there were snapped trees lying in every direction. It was like day one all over again. Our stomachs ached for lunch before we'd even found our plot. Our third attempt was a keeper, and it landed in what we had called "mid-range mortality" from the boat—meaning the trees hadn't died very recently or as long ago as the telephone-pole skeletons farther south. The forests in these areas had been responding to the death of the cedars for a couple decades, and what we discovered that day was utter chaos. The brush density was so thick that Kate could hardly position herself to take accurate height measurements. Vibrant green engulfed the dead trees, and it was easy to lose sight of one another in just moments. We kept track of each other's whereabouts by the sound of rain jackets rubbing on conifer needles and measurements hollered out, assessing the big trees first. There were so many young, small ones to measure that we had twice as much to do as ever before. Hummingbirds hovered around our pink flagging, the color of succulent flowers. *Were the hummingbirds in previous plots?* I didn't recall seeing them, but then, I wasn't looking for them, either. They seemed to surprise me, like something new.

I pushed through brush to get a better view of a dead yellow-cedar tree and stumbled into Odin. He was crouched beneath the small conifers with the "Pojar," our one reference book for plants.[2] Odin held it open with both hands, and I could see a stem with small white flowers resting in the crack of the spine. He'd given up on his jacket for rain protection and was kneeling on it instead. Leaves and needles were stuck all over his drenched shirt.

"Got a few species I haven't seen," he said. "There's a lot more going on here in the understory than the usual." He looked up at me standing with my clipboard in the rain. "It's very slow going," he added.

"Also slow going up here in tree world," I said.

I felt cold and frustrated by the new set of challenges for us all. But as the work hours wore into evening, the scientific reality of the plot became more and more intriguing. The poking and prodding, the double-measuring and double-counting for accuracy, the extra measurements for grasses and shrubs, and flowering plants growing amidst mosses—every bit of it pointed toward a story of survival amidst loss and death. It was a forest letting go of what was and becoming something new. Some members of the community were finding ways to make the best of their shifting surroundings.

Plot seven was similar to six. Kate said working in the midrange forests made her feel claustrophobic. The crowded shrubs pulled at our clothing. We crouched under saplings, making barely legible notes. When we reached the beach each night on the other side of the wall, we swung our arms in circles, intoxicated by the feeling of freedom after the tight grasp of the green throughout the day. My fingers were wrinkled like I'd been soaking for hours in a bathtub.

"It's like these forests are in midlife crisis," I said as we carried our kayaks down to the water after plot seven. This early on, there was no way to know how the chaos would play out—which species would persevere and thrive over time. But the dense green that nearly swallowed us on the hardest days in the midrange was enough to set my hook in further. If we measured enough plots in the sites to the south, where more time had passed since the cedars had died, I believed we'd discover the new forests taking shape. I'd be able to tell what happens over time, to find out what species grow and live on in the changing climate.

We didn't get ten plots completed that first trip. On day twelve, seven plots in, we ran short on food. That night around the campfire, P-Fisch and Odin added more butter and Snickers to our restock list. Kate and I added neoprene gators—wrist cuffs that fishermen use to help keep the water from dripping down their sleeves. I got the satellite phone to work from the beach and confirmed a pickup for the next day.

We waited on the beach in the morning for the floatplane to arrive. One Clif Bar remained, my last bit of food. I promised myself I wouldn't eat it until I saw the plane land. From my field journal on July 8, 2011:

> Dry bags packed and sitting on the beach. . . . Odin up in the trees reading about the Chukchi Sea. Kate cuddled up in her raingear in fetal position, spooning the roots of an old western hemlock—*Tsuga heterophylla*. P-Fisch, head in hand, pondering over his waterproof journal. Me, warm in my "sleep long underwear" for the first morning in days, weeks

now, one sacred pair I've kept dry for the tent, and I'm sacrificing them today in the hopes we get out.

I hope I didn't jinx us.

My fingers hurt writing. My hands are covered in scrapes and little cuts and devil's club sticks. Nails and fingertips actually ache from all the climbing up, under, over trees and shrubs. The fog is standing strong today, maybe too strong to fly. Thick layer resting on the Khaz peak. We watch the soup lift, then sink back deeper into Slocum Arm.

We're waiting. We're hungry. We're already delayed. The two days of extra food in case of stranding is gone. Sitka seems like a fantasy. Visions of drying out, recharging, talking to loved ones. Tearing into salmon, fresh greens, protein and fat. Seven plots down and 33 to go. . . . We are constantly working to stay warm, to access our sites, paddling, hiking, scrambling, hanging bear bags, hollering, measuring. . . . Middle of the night, waking up to hunger pangs. I'll take that spoonful of oil in my oatmeal, yes please. California feels so far away, and I think we've all forgotten what sun feels like upon our skin.

Solid cement startled me when I stepped off the floatplane in Sitka. After expecting to tilt or trip with every step, I couldn't walk a straight line with a decisive gait anymore. We hadn't showered in two weeks, but we staggered from the dock to the airport café, collapsed in chairs at a table, and ordered omelets and pancakes. We had sixty hours until we'd load up and fly back to Slocum, and we needed half of those hours for sleeping. We'd have to find a way to get two plots a day in the trips ahead of us, and I had no idea how to make that happen.

The waitress served steaming-hot plates of fresh food. P-Fisch dove in immediately. Kate sat there for a moment staring at her golden omelet, and then burst into tears.

"Are you crying?!" I asked.

"Geeeeez," she said, wiping her eyes and turning red. Everyone laughed. "I never said a word about how hard it was out there. Come on, give me this moment. Yes, I'm crying!"

Odin, P-Fisch, and I fell quiet. I plugged my phone into the wall for a charge and turned it on. Hundreds of emails began downloading. It shook and beeped with every delayed text.

"Umm, I think town is gonna be really weird," Odin said.

It felt as if we existed between two worlds. The one outside of Alaska flooded back in megabytes and messages, while the outer coast kept its firm grip. There wasn't enough space in those limited hours for both of them. I scanned the messages for the ones from family, friends, and the fellow I missed. I made two calls that day—one to my mother and another to Jonathan. Both left me feeling a great divide between my life amidst the trees and any "normal" life as I listened to them describe the days that had passed. Then I stopped making calls and focused, instead, on doing everything possible to prepare for another round. Basecamp #2 on the outer coast was waiting. Ultimately, what happens next—what species were benefiting from the changes out there—would be the link between those two disparate worlds.

CHAPTER 5

Countdown

DRESSED LIKE FISHERMEN in waders, we huddled around the camp stove counting the minutes until the couscous slathered in rehydrated cheese sauce was ready to devour. Odin was singing Johnny Cash's "Folsom Prison Blues," swapping "Slocum Arm" into the "stuck in Folsom Prison" line.

"I ain't seen the sunshine since I don't know when!" Kate, P-Fisch, and I sang into the rain. August 2—we were 1,216 trees and 552 saplings in, with 27 plots completed, 13 to go. Odin had lost more than a dozen pounds from his brawny frame. I had added a belt to my rubber pants repertoire, which conveniently made my bear spray more accessible while controlling the butt sag.

Rain we could tolerate. It was the blustery storms we needed to avoid—the same ones that could restructure a forest. Forests, however, are both dynamic and resilient. Gusts of wind tear down trees, creating patches of blowdown on the landscape like pickup sticks on the floor. Saplings grow into the gap. And when a centuries-old or millennia-old titan rots from the inside out and eventually topples, it opens up the canopy for light to reach below, for seedlings to take hold on decaying wood, for life to begin again.

Part of what makes an old-growth forest so resilient is what scientists call the "reverse J."

I first learned the term in Kevin's class. I'd been searching for it all summer—a sign that perhaps the yellow-cedar population wasn't so doomed after all.

In 1898, a French forester, François de Liocourt, published a paper that revealed the structure of a natural forest.[1] He wanted to determine the maximum financial yield that could be obtained over time by selecting trees of certain sizes in cycles for lumber while leaving the others to grow. What he found was a common structure across the various fir forests of northeastern France and a mathematical equation that could explain the shape of a forest as far away as Southeast Alaska. In a healthy old-growth forest, or a forest composed of trees of various ages, the relationship between the number of trees and the size-classes—groupings of tree diameters—remains constant.[2] De Liocourt's discovery offered an early target for what foresters would later call "sustainable" forestry, because a flourishing population of young individuals could eventually replace the few old and mighty. The shape of the curve—the reverse J—shows there's a young generation to grow into the old. A healthy forest needs the vernal green in the shade of those who die, like children running around the knees of their grandparents.

Back at UC Berkeley, Kevin had drawn the arcing line on the chalkboard and described thousands of seedlings at one end and few giants on the other. Scientifically, the concept and the logic behind it made sense to me. Over the course of any individual life, a tree faces challenges: competition with others, limited space, disease or injury. Few survive to become elders. It wasn't until our third trip to Chichagof that I really understood what a reverse-J distribution—or the lack thereof—meant for yellow-cedar.

P-Fisch hovered over the waterproof map and punched the GPS button to calculate how far we'd have to travel in the morning for another site. I crawled onto the tallest rock between our cook spot and the lapping tide and raised the radio high in the air with the volume up full blast. Cold water slipped down the sleeve of my moldy jacket. I stared down at four equal portions of butter slices stacked carefully on our one cutting board. Rain slipped off their oily, yellow surfaces, and my mouth started to water.

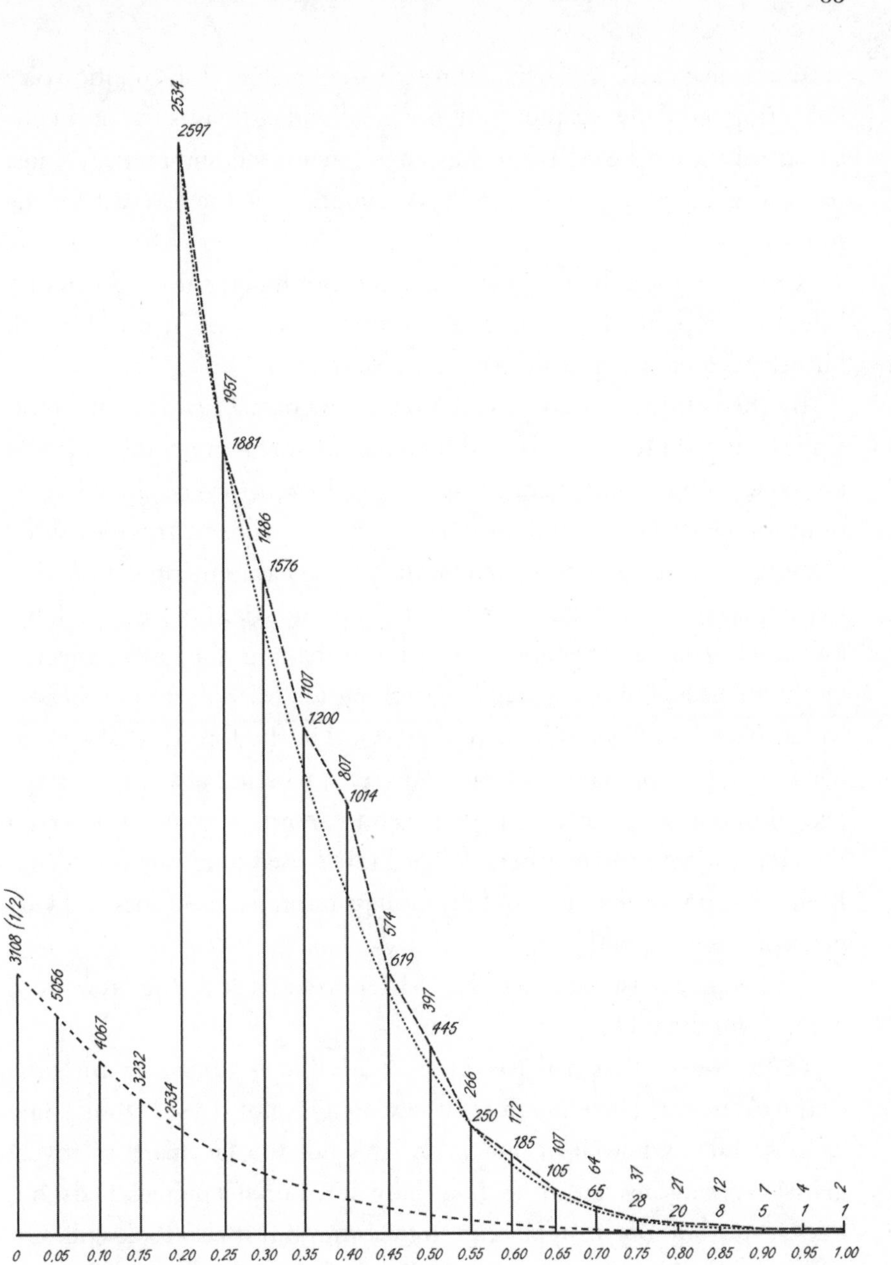

The reverse-J curve from de Liocourt's study, showing tree diameter on the x-axis and counts on the y-axis. Reprinted from François Liocourt, "De l'aménagement des sapinières," *Bulletin Trimestriel*, Société Forestière de Franche-Comté et Belfort (Julliet 1898): 396–409. © Crown Copyright, Forest Research.

"Seas, five feet," we heard through the crackle. The robotic voice delivering our daily weather from my handheld radio was the only consistent one we'd heard from the outside world all summer. "Winds, ten knots, increasing to fifteen by afternoon," was the forecast for the morning.

This time, the ache in my stomach was not hunger alone. I couldn't bear the thought of spending the next day—August 3, my thirtieth birthday—measuring dead tree after dead tree.

By the third trip, I could pick up a soggy datasheet from any plot, scan the list of observations, and then see the forest in my mind. It was a privilege that office scientists rarely get. I could review the observations for a specific tree and know the texture of its bark, and remember something more about its character than our measurements captured—the ridges of its roots crawling into the ground, the limb it lost down low, the five-leaved creeping raspberry at its base. In the forests affected by the dieback, I was constantly searching for patterns and outliers—which trees had died and where they stood in the forest, what species lived on and grew vigorously tall and green, whether any surprises appeared in one stand that we hadn't seen in another. Trends I observed I'd later test with the numbers. What I saw I tried to remember. What I felt—the physical strains and my earliest impressions of loss and vulnerability—stuck with me.

"We've got reasonable weather," I reported back to the crew from my spot on the rock.

In the weeks that had passed since Maddog's tears over the omelet, we'd gotten fast enough for a few double-plot days. It took some luck—winds and swell in our favor, sites not too far apart, relatively manageable hiking terrain without blowdown or steep creek beds, big trees instead of lots of small ones to measure. But it also took skill. On successful double days, we worked in a carefully orchestrated dance, no longer stumbling or stalling, but measuring fluidly, almost gracefully at times.

Plots in the midrange mortality added challenges to achieving a double day. With so many small trees and shrubs extending toward the light, they remained the most demanding. But the sites where more time had passed since the death of the yellow-cedar trees—they were more manageable. It was as if the forest had worked through its chaos. Some species prospered where others dwindled. Western hemlock trees stood tall next to the dead cedars, the latter fallen prey to decay, their barren bodies exposed. The live, full branches of the hemlocks extended around their trunks, as if the young grew up hugging the deceased. We walked underneath canopies being reshaped by hemlocks, their needles filling the space where the feathery cedar foliage had once been.

Kate dumped dehydrated cheese sauce into the pot. Her subtle grin, coupled with the vigorous stirring, let us all know the feast time was close. "So, here are the options," P-Fisch said. He rubbed his brown-bearded chin and adjusted the pencil secured under his red baseball hat. "There's a site farther south that we could try. Three miles by kayak to the landing spot. Then the random distance we have to hike into the forest is half a mile, if we hit it right. By the site classification, it looks like it'll be pretty thick with regeneration. It could be slow going just to get there before we even start our work."

Our toughest hike to date had taken us an hour of scrambling through a tangle of forest only to cross the length of a football field. I often wished I could be a monkey or a mountain goat and sometimes both at once.

"What else have you got?" I asked, as Kate scooped sloppy cheese pasta into our bowls. Odin divvied up the salted butter stacks.

"We've got a healthy one to the north," P-Fisch replied, after pressing a few buttons on the GPS. "It's not as far to kayak, and it should be a cakewalk to the site without all the dead." He paused, and then suggested, "Maybe we should save the easy one for a windy day."

I weighed the risks. If a storm hit, he was right; we'd need that site to keep pace with at least one plot per day.

"Your birthday. Your call," P-Fisch concluded. "We'll get them either way. We've made it this far, and we've got some more power coming."

We were getting down to the wire. Paul Hennon was scheduled to fly in the next afternoon with two volunteers from the Forestry Sciences Laboratory out of Juneau. They'd arrive with a small inflatable boat, called a Zodiak, and use a motor to get around Slocum. For a few days, they'd help locate stands and set up flagging around our plots, trimming our search and setup time.

I turned the radio dial to silence the crackle and looked out across the foggy inlet. "Okay, let's go with the healthy one," I decided. I didn't want to spend my birthday surrounded by dead yellow-cedar trees, pushing through saplings and brush and the devil's club (a species aptly named *Oplopanax horridus* for its horrid prickly spines). The decision meant gambling on the weather in order to reach the other site another day, but spending a day surrounded by live yellow-cedar was the birthday present I could give myself.

I wondered what my mother was doing that day, and what filled my father's days, weeks, and months of late. After dinner, I snuck away to use a few precious minutes on the satellite phone and checked the voicemails on my personal cell. There were birthday messages from my mother, from my brother and his wife, and from friends back in California. Nothing from Jonathan. I missed him; he should have been on the plane with Paul, flying in for a few days to help us out. I'd been counting down the days, aching for his fun company and a reconnection to my life beyond Slocum. For nights, I'd slept with my head resting on a t-shirt he'd sent me in Sitka; it had arrived in a Ziploc bag, and when I opened the seal, his sweet smell seeped out.

"What do you mean 'you can't do *this thing*'?" I had asked him in a conversation over satellite phone around the twentieth plot—just one trip and one restock prior. The time delay by satellite made it impossible to discern whether the long silences were ours or the technology's. "Like come out with the volunteers to the coast?" I clarified.

"This thing." Long pause. "Like us." My heart sank. *Why? Because I was gone for so long? Is there somebody else? I'm coming back. I promise.* I had so much to say but couldn't say any of it. I looked up and there was a brown bear approaching me.

"I've gotta go. There's a bear," I said, raising my arms to the sky with the satellite phone in my right hand.

"Hey bear. Heyyyyy bear."

I held my ground and avoided eye contact. The bear stood up on its hind feet, raised its massive snout, and snarled. Odin, Kate, and P-Fisch hollered from across the beach. The bear stepped toward me, then turned away, slipping into the forest. That night, before crawling into our nests, we hollered and banged pots and pans in the forests surrounding camp, attempting to define our territory. It worked.

Jonathan was supposed to be flying here tomorrow, I thought. The idea still hurt.

Odin passed me a bowl of butter-coated Cheesy Mexi, our best dinner of the limited options—couscous, quinoa, or pasta with either tomato, cheese, or curry sauce from powder.

I clutched the warm aluminum in my hands, and then said, "I want to be in the green."

"Yessss!" Kate agreed.

"And let's sleep in half an hour later," I said.

"Everyone wins on birthday day," P-Fisch declared.

"Just wait until the whiskey," Odin added. He dug deep down into our metal bear box and pulled out a buried bottle.

I hoped a day in the healthy forests would stop me from thinking about dead trees and rising temperatures and the love I'd left behind and lost. I thought maybe for one day I could celebrate the surrounding beauty: humpback whales offshore and eagles soaring, live yellow-cedars with bowed limbs swaying in the wind.

"I think we should have a climate-change-free day," I said.

"Umm, what do you mean?" Kate asked. "It's, ahh, a little late for that."

P-Fisch laughed.

"Like for twenty-four hours we don't talk about it. We just try to be in a beautiful forest. Is that even possible?" I said.

"Huh," Odin grumbled, scratching the itchy devil's-club pricks on his forearm. While Kate, P-Fisch, and I worked up high in tree world, craning our necks in the plots, Odin always measured small plants down low—some only centimeters tall. His hands were puffy and worn from sifting through the green to count seedlings.

"Let's try," he said, and pointed to a sea lion staring at us.

This was the third trip, and we were camped on an island near the northern end of Slocum Arm. It was a small island—probably only a few acres in size—but we called it "our island" and it felt like our island. We'd walked the perimeter and crossed the interior when we first arrived, checking for bears. With a false sense of security, everyone said they slept better on our island than they did in the other camp where I'd encountered the bear. Only the occasional "pfooof" sound of sea lions surfacing for air broke the eerie quiet in the night.

THE FOLLOWING MORNING's hike was easy going, and not because we'd gotten stronger and more agile. We didn't need to dodge dense blueberry and menziesia shrubs, or navigate between crowded spruce and hemlock saplings. Thick moss formed a soft carpet beneath my feet.

I led. We didn't talk much. We moved quickly. I wove between trees with the GPS unit in my hand, following the little arrow until we landed at a new plot center.

I'd expected to work beneath a closed canopy of an old-growth forest that day, under the umbrella of yellow-cedar foliage. I'd expected to record measurements for yellow-cedar saplings and more seedling counts than usual. In my journal from that day, I wrote, "I planned for us to measure what I thought would be our biggest and best . . ."

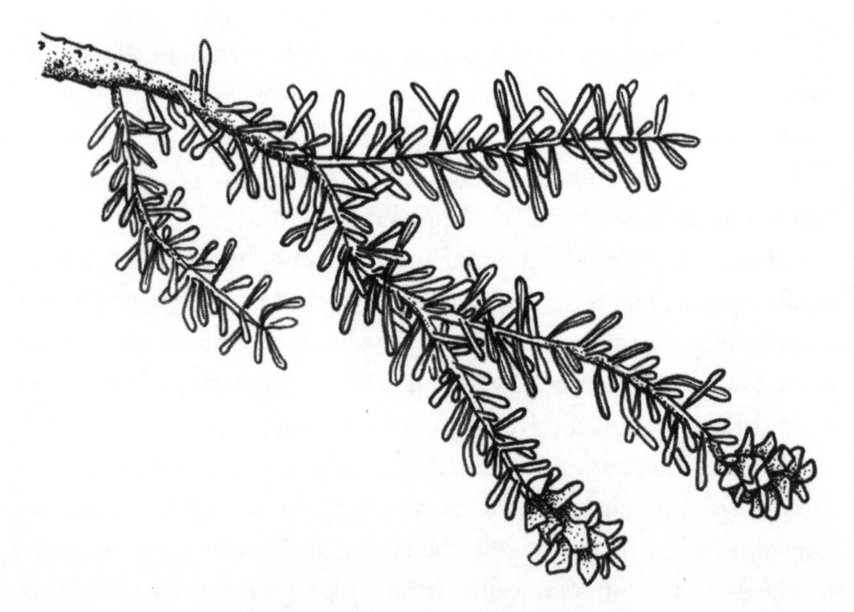

Tsuga heterophylla (western hemlock).

I expected to find the reverse-J curve—that structure of a healthy population with many small trees reaching up toward the old. We didn't find it there.

"Yellow-cedar. Tag 972," P-Fisch called out. He hammered a nail through a hole in a metal tag about the size of a quarter.

"59.6 in diameter. Alive," he reported next.

I looked up high at the foliage above. It was green, but there wasn't much of it. "Ten percent," I said. "Alive but not so happy."

We moved through the measurements for a hemlock tree beside it, then on to the next.

"Yellow-cedar. Tag 974." He hammered again, metal on metal, the blows echoing.

"51.8 in diameter. Alive."

"Forty percent," I said. "It's only got about 40 percent of its crown. This one looks better but still not great."

Even the "healthy" yellow-cedar plots weren't that healthy after all. *We agreed—no climate change talk*, I thought, and kept working.

I placed my tripod in the center of the forest plot to take the photograph that I would later use to estimate light. Over the summer I had already taken nearly thirty of these images with dead trees dominating the views. I looked up at the green canopy, soaking in 180 degrees of old-growth forest through the fish-eye. I fired the shutter multiple times, then looked up at the canopy with my own two eyes.

I closed them and paused in the dark.

When I opened my eyes again, it was as if the forest was flashing forward in time. I imagined the lingering leaves of the yellow-cedar trees turning from green to yellow to rust and then brown, falling to the floor at my feet. I imagined light cracking through, and above my head, leafless branches reaching across the sky. Then I could see the hemlock, standing tall after decades had passed, hugging the skeletons of dead cedars.

The rain held off. We finished early and lay down on the moss. I was confused by the data we'd recorded. The stand had appeared healthy from the boat survey, but our experience inside the forest, and the numbers in the plot, showed something else was going on. Odin had counted tiny yellow-cedar seedlings in his quadrats, but fewer than I had expected. There were some dead yellow-cedar trees too.

Maybe those dead trees are normal deaths—background mortality, I thought.

Tree death is a part of life in any forest, just as human death is in a given population. What demographers depict in a population pyramid is similar to what ecologists see in forest structure. De Liocourt's calculations helped create a method for determining the expected rates of tree death in a forest of trees that spanned across ages. So "normal" yellow-cedar deaths would be deaths in the absence of climate change.

There were live yellow-cedar trees in the plot—some big ones—but their crowns were far from what any scientist would classify as full and vigorous. Something wasn't "normal."

Is it spreading? How long do these trees have left?

"No yellow-cedar saplings?" I asked, noticing the counts were all western hemlock and spruce. Kate shifted into fetal position and curled up like a kitten ready for a nap. Odin thumbed through the Pojar plant book.

"No," P-Fisch confirmed. "We walked through a couple on the way in," he said, "but there weren't any inside." He was sitting up, sharpening his pencil with a knife. I kept staring at the data.

"I know, kind of weird," he added.

"Could be deer," I said. Black-tailed deer were notorious for munching on the tender yellow-cedar saplings. "But if the big trees are stressed or maybe even dying, and there aren't enough little ones—" I stopped myself midsentence.

We agreed—no climate change talk.

"Do you hear a plane?" Odin asked, shoving the measuring tapes into his pack. I pulled my hood down to listen. It was approaching from inland, the route pilots took on relatively clear days. I aligned the datasheets carefully in a stack and put them inside a protective metal folder.

"Paul and company," Kate said, bolting upright. "They're gonna make it in. Let's get out of here."

As far as science goes, I was convinced that the chronosequence would work, that what I could see happening over time—a story of death and rebirth—was getting captured in the numbers. But once we finished on Chichagof Island, I'd still need to measure the healthy sites farther north. If the reverse J for yellow-cedar existed anywhere in the transect, I was sure it would be there, on the outer coast of Glacier Bay, where snow fell in winter and likely lingered longer in spring. Like a community of people whose relationships all change after the death of an elder, I could see the same holding true for the plants. It was a new kind of dynamic forest under a changing climate, with species responding to one another and to the shifting conditions.

I could also see there weren't enough young trees to replace the old. In the forests where climate change was taking its toll, the future for the cypress looked bleak. Biologically, the other plants taking over were the ones that could make the best of the new environment surrounding them. The forest would live on as the shape of its community changed. Yet, as I wrote in my journal, "I tend to sense loss" amidst the dead cedars. It was a loss for the yellow-cedar, I noted, not a joy for the hemlock growing up. The human response—mine included—hinged more on values. What do we use? What do we need? What do we love? What can we fight for and what must we let go of?

Why did it matter if yellow-cedar turned over to hemlock, or moss gave way to grass? If oak ferns waned without as much shade and blueberry branches climbed toward the light?

They were questions I couldn't answer myself. I was far from done with my plot measurements, but I was already thinking about what I would do next. The second half of my study began to take shape in my mind.

We paddled south back to our island, looking across Slocum Arm at the head of the Khaz. Kate set a fast clip from the front. Marbled murrelets skimmed across the water like skipping stones. We'd seldom seen them farther south amidst the dead cedars.

We had twelve more plots to go, then another set of ten in the National Park to the north—a year later—to complete the chronosequence, but that day, my birthday, shifted my attention from plants to people.

Greg Streveler had been steps ahead of me the whole time. "Cedars are the quintessence of conservatism," he wrote me, not long after we first met outside the woodshed in Gustavus. "They grow slowly, reproduce sparingly, and hold tenaciously to their spots, once won, with an array of chemical and physical defenses. Of course, when they are forced aside, the more exuberant members of the choir stand up and shout hallelujah! So, what is the value of cedars, if they hold down that exuberance in the few spots they manage to preempt? Perhaps the same as a very old and wise person does in the society of people."

My data would show which species were those "exuberant members of the choir" standing up to take their place. But why that mattered could only come from Alaskans I'd yet to meet, from loggers, naturalists, Native weavers, and many others who used and valued these yellow-cedar trees—the people who knew them best.

Holding steady in our strokes, we moved toward camp from plot #28, the bow of the Whale breaking through whitecaps forming in the headwind. I began a new mental list of interview questions as I fell into the rhythm of paddling.

When you think about yellow-cedar trees dying, how do you feel about these changes?

Has the death of these trees changed your use of the forest?

We hit a patch of bull kelp, the hull making a loud clunk against their long, snaking bodies—giant seaweed whips with a bulb at one end. I pulled the cord for the rudder, lifting it out of the water to free us forward.

Is this dieback a story of loss for you, only loss, or is it also one of opportunity?

When our kayaks touched the shore of our island, Paul Hennon was waiting with a twenty-four-pack of Rainier at his feet. Avery, the same pilot from our aerial survey, had left it as a present.

"Avery said you were the only woman wearing a skirt that he'd ever dropped off on Chichagof." (I'd taken to wearing a nylon hiking skirt over my wool long johns for the days we set up basecamp. Underneath rubber pants, I was always soaked in sweat, and the freedom of the skirt made carrying gear over rocks and beach logs easier.)

"You look a little more weathered than when I saw you last," he added. I stepped out of the Whale and pulled my lifejacket over my head.

"I'll take that as a compliment," I said, giving him a big hug.

The seven of us circled around a smoldering fire that evening in the fog and drizzle. We slopped pudding (made from powder) over homemade cookies Paul's wife had baked. We told stories of the good and bad—our first double-plot day, the trip when we'd run out of food, the morning we'd inadvertently crawled into an abandoned bear den, that spruce tree nearly six feet in diameter. Odin recalled when the Impulse fogged up so much that we couldn't measure trees until the sun returned. P-Fisch remembered the orca offshore. Kate recalled a restless night, dreaming in paranoia that everything she touched inside the tent was wet. Odin poured a drop of whiskey on the beach before we drank—an offering to Earth and a tradition he'd first witnessed in Siberia.

I fell asleep that night feeling more satiated in my stomach than I had in weeks, but lonelier than ever before in my life. The breakup, the fact that Jonathan wasn't on the plane with Paul—it all felt like more symptoms of science and the dead trees taking over my life—and for what, I wasn't sure yet.

"TODAY WE CONQUERED the impossible," Maddog wrote in her field journal on August 17. "We finished 40 plots! . . . Odin did his last 8 understory plots. . . . Paul took his last DBH and nailed in the last tree tag. . . . Odin and Lauren took their last cores, and I did my last Impulse height measurements. . . . We had done it, good god, we had done it."

Captain Charlie picked us up precisely on schedule, and we rode the swell on the outer coast of Chichagof one last time that summer.

From the stern of the boat, I stared at the head of the Khaz. I watched it disappear in the distance, like a curtain dropping shut at the end of the play—our play. We had mastered our parts, given everything we had, and then, just like that, it was over.

Reentry was tough. I needed to recover. Yet part of me wanted to go straight back to the woods. Somewhere out there, I'd let go of the life and love I'd known in California. I missed my family and friends, but there was a welcome simplicity in days reduced to survival and science.

Back in Sitka, we spent mornings sleeping and afternoons cleaning gear. I kept quiet, regrouping. A doctor at Stanford would later tell me that what followed was like a form of "posttraumatic growth"—a psychological shift in thinking and relating to the world that results from experiencing significant challenges.

On August 23, Kate and I loaded up Captain Charlie's truck, and we picked up the boys at the Forest Service bunkhouse on our way to the ferry terminal. I boarded the ship carrying a backpack full of tree cores, the smell of cedar trailing behind me in the wind. I rolled out my sleeping bag and pad on the rooftop solarium. The engine hummed. The floor vibrated gently beneath me. Cold air from the Inside Passage chilled my exposed face. Inside the warmth of my cocoon, I fell asleep with my hand on my stomach. In my journal, before we arrived in Juneau, I wrote:

You are in the belly of what's wild.
The womb of Wilderness that remains.

I staggered, drained and dreamy, off the ferry. *What now?*

PART II:
BIRDSONG

If what a tree or bush does is lost on you,
You are surely lost. Stand still. The forest knows
Where you are. You must let it find you.

—David Wagoner

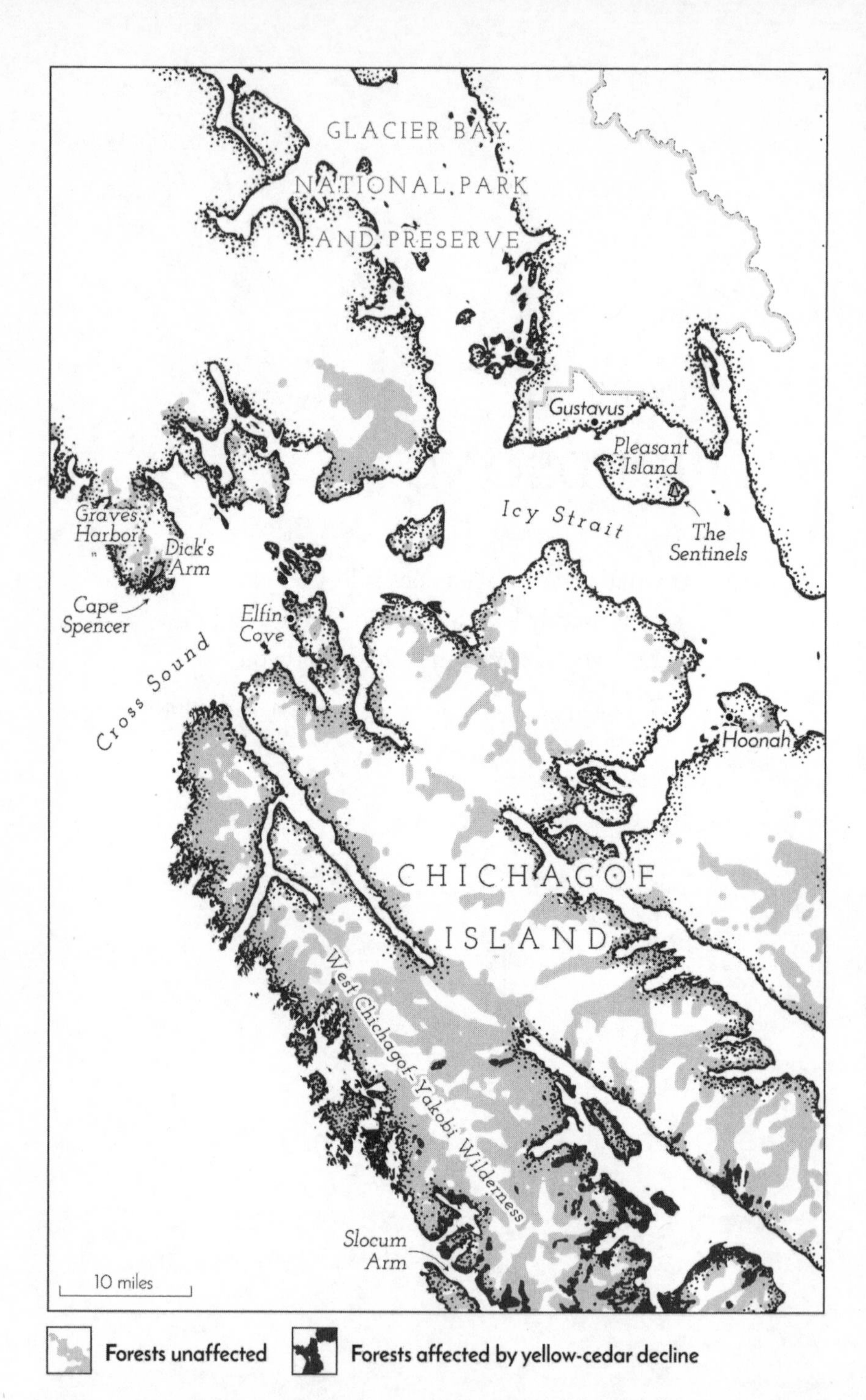

Detailed map for locations referenced in Parts II and III. Aerial surveys I conducted with researchers from the United States Forest Service revealed relatively healthy forests on northern Chichagof Island and in Glacier Bay National Park and Preserve. So this map shows a lot more unaffected areas than affected areas, as opposed to those depicted in the map for Part I.

CHAPTER 6

Thrive

THE END OF the play wasn't really the end, only a ten-month intermission between acts. I spent those months in California; they began lonely and became monotonous. I'd mailed Jonathan a letter from Sitka, asking him to pack my things into my car and return my car to my home. My little blue Subaru was there, dusty and covered in leaves, when I arrived, and it started slowly, still choking up months of a life laid fallow. I kept busy and filled the fall with everything not Alaska—fresh greens and bike rides in the sun, long nights under a dry comforter, and cotton clothes. The forests had demanded all my physical strength; now the data required mental stamina. Even with the help of a team of research assistants, it took months just to enter every measurement into a digital database. Then it took months more to test the patterns I'd seen on Chichagof, to run the stats.

In the end, after 40 plots—240 quadrats on the forest floor, more than 800 small saplings and 1,700 trees above—I still needed the forests on the coast near Glacier Bay, at least ten more sites. By March, I was only halfway present in California, split between data crunching and designing my interview protocol in one place and plotting field logistics in another. I planned for Act II to take place in the National Park, with an encore on Chichagof with my crew to download data and replace batteries in the temperature devices I'd left out there all year;

then I'd stay without my crew for a few weeks to begin interviewing. I was eager to finish measuring plants and start listening to people.

So we did it all again ten months later, when the weather window opened in June. There was mostly the same cast of characters, with a few understudies subbing on the new stage. P-Fisch said he was determined to hug (and measure the diameter of) absolutely every tree in my study. Maddog said she felt she'd left the chronosequence incomplete. She needed to see a thriving population of yellow-cedar and agreed that the outer coast of Glacier Bay was our shot. Odin wanted to return as well, but he'd gone back to graduate school after Chichagof and reserved the summer for his own fieldwork. A quiet, aspiring botanist named Tomas (pronounced *toh-MAHS*, not *TA-muhs*) took his place. We'd work together in two inlets: Dick's Arm and Graves Harbor.

Captain Zach on the FV *Taurus* agreed to run his boat for my second survey of the coastline, this time farther north. Instead of the Khaz, we faced Cape Spencer—a jumble of rocks jutting into Cross Sound, a place where tides rip and winds stack waves in another tumultuous meeting between the archipelago and the blue horizon. We would only round the point at slack water, when the tide was neither coming nor going, but catching its breath before the next race.

ON JUNE 22, 2012, the night before we left for the coast, a woman named Lori Trummer hosted us all at her home in Gustavus, the gateway to the National Park. She was a friend of Paul Hennon's, and Paul was joining us again for the first few days. Greg Streveler came for dinner. We talked over pasta while my crew consumed enormous portions, anticipating another expedition for what we'd affectionately termed "the outer coast fat camp." Paul had printed a map of sites he'd sampled throughout the archipelago for a genetics study on yellow-cedar. A grid overlaid the map's colors and contours, and any box with an X in it had already been sampled. Nearly all the ones shaded yellow for cedar had Xs on them, except where we were headed in the park. So he, Tomas, and another colleague from the Forest Service would

Yellow-cedar on the edge of muskeg near Graves Harbor.

collect foliage for a day, while I ran another boat survey with Captain Zach and the rest of my team.

Just the week prior, Paul and I and another researcher, Dustin, had flown the coast from Cape Spencer north, soaring inland across forested lands and glaciers within the park boundaries. We combed 3.3 million acres for yellow-cedar, looking for healthy trees, dead trees, and dying trees—any signs of stress.

We might just find the reverse J after all, I thought, wondering what the forests we mapped from above would look like from the inside. Following the meandering lines of the glacial moraines, I relaxed my eyes when we flew over the ice. Whenever we hit green again, I squinted, searching with Paul for the distinguishing shape of cedars.

"It's gonna be cold out there," I said through the headphone system. "Definitely chillier than Chichagof."

We flew as low as we could safely and recorded observations in a real-time tracking system. We created the first detailed map of the population from aerial surveys near its northern limits.

A Park Service ecologist had told us that Greg was probably the only person who knew enough of the landscape to compare what we'd mapped over two days of flying with other observations from flights, fieldwork, and historical documents. He'd also be the one to tell me about the local winds and tides, and any potential locations for base-camps, in the inlets where I'd received permits to work. Greg showed up at Lori's wearing wool pants with suspenders again and mulled over the map between bites. He asked a few questions, then confirmed areas where he'd seen yellow-cedar trees, and other areas he suspected.

With an audience there—Paul Hennon, Lori and her partner, and my hungry crew—Greg and I stuck to ecology and logistics. No philosophy. No questions about meaning. It seemed respectful with everyone's focus on the trees, easier that way perhaps. He played along, drawing from years of his own geological research on the coast, as if science was all that mattered.

I fidgeted with the napkin in my lap, thinking back to his probing questions about yellow-cedar as my muse, about what I would do after answering how the forests develop following the death of cedars, if I would just be another scientist monitoring another species to extinction.

I had questions for him. They were more formal now, worded carefully for the next phase of my research. Between crunching numbers on plants, I'd spent months researching and formulating an official interview protocol.

Has the dieback altered your use of these forests?

Are there forests with yellow-cedar that are important to you because of what you gain from them?

Are these gains different from healthy forests versus those affected by the dieback?

I wasn't going to ask any of these there at dinner, but I'd already decided that if Greg was willing, I wanted him to be one of the many people I'd interview for "the pilot phase." Similar in practice to the week I'd spent with Paul and Kevin on Chichagof bumbling through methods, the pilot stage for the social scientist provides a chance to identify what works and what doesn't, to make revisions before spending months committed to asking the same questions. A term like "forest dieback" can mean one thing to a researcher or scholar and an entirely different thing to someone living near a forest. So piloting is not only about refining the issues addressed and the types of questions asked, but also about *how* they're asked. It's an effort to find common ground in language, a process that would ultimately enable me to identify patterns in hours of recorded conversations.

I was shooting for about forty-five or fifty interviews. I would come back to the archipelago the following spring to conduct them with people I'd yet to identify. But by the time I was sitting at the dinner table with Greg, I knew I'd talk with residents living in towns near remote forests with live yellow-cedar trees and patches of dead ones—so that I could talk to them about both. I also knew I'd search for people who represented a range of uses of the forests. Scientists call this strategy "intensity sampling"—selecting information-rich cases of the phenomenon of interest—and it meant I'd talk with loggers who depended on cedar in their businesses—Native carvers shaping cedar trunks and weavers using their bark, forest managers plotting timber sales and conservation strategies, deer hunters, and scientists and naturalists, like Greg.[1] I expected that some people would value the trees for their wood—for money or direct uses that came with harvesting; I figured others would value their existence—alive, in place—for reasons unknown to me then. My plan was to find these people by what's called "snowball sampling." One person refers you to the next, and the next refers you to another, and so on, until your snowball is rolling down the hill, getting bigger and bigger. You arrive, eventually, with a portrait of many perspectives.

Greg took the map with the Xs all over the archipelago and turned it over. He drew an oval and labeled it "Island," and then marked the coastline forming the inlet where I was headed, "Graves Harbor."

"Here's a good place to camp," he said, drawing an arrow. Then he drew another circle, gave it some fur in a flurry of lines, and wrote "Bad Bear" beside it. On reconnaissance for an archaeological study, he had just returned from the same place that we were going.

"There's a territorial bear, a large one. It was kind of limping around, pretty angry."

"Noted," I said, taking back the map. *Ten more plots to go, and then I'll come back. We'll talk, really talk. I promise.*

An elephant was sitting between Greg and me as we all talked about plant ecology, as if that was all that mattered, and the only thing I could say to acknowledge it before Greg left was a commitment to return.

Looking unsatisfied, Greg got up from the table when conversation dwindled.

"I'll see you in September, if not sooner," I told him.

"I'll be here," he confirmed. He left before dessert.

"Let's pray for sun," P-Fisch said before we all went to sleep that night.

"No more praying for sun this time," I said. "It brings those westerly winds, and Graves is less protected than Slocum. If you're praying for anything, pray for rain and fog, but not too much of either. We need the in-betweens."

"Okay, pray for calm then. Smooth waters," P-Fisch clarified. I rolled out my sleeping bag on the floor.

"No sun!" I said.

MADDOG WROTE IN her journal the next day that it was "as if we never left" the outer coast. We fell into the same routine effortlessly: "Unload boat, pick camp spot, find food cache, set up tents, tarp, hang bear bag." We had a few upgrades for Act II—a pump-action shotgun for bear defense, a far better choice than the bolt-action rifle for the even larger bears we could encounter, and an electric bear fence, which Tomas set up carefully around our tents.

"This really works?!" he said, holding up a handful of lightweight plastic poles and what looked like a spool of white string. It was electrical wire coated in plastic, which we'd connect to a couple of D-cell batteries.

"This pretty much looks like dental floss for a bear," Maddog declared, grabbing the spool. She inspected the squatty batteries.

"Really?" she said. "I use these things in dinky flashlights. Come on, now. For a bear?!" We'd already seen a mother with two cubs on our boat ride in.

"I was skeptical too," I admitted. "But I watched a few YouTube videos of grizzlies just wandering around the things with stashes of food sitting right inside. The bear biologists swear by them."

"Maybe we'll sleep better," she said. "You know—like peace of mind. Whatever, I'm in." She passed the spool to Tomas, and he began marking the perimeter of our territory with stakes for the wire.

When we walked into our first plot in Graves Harbor, I thought of John Caouette—my friend in the rain outside the Paradise Café, the forest statistician, the man who'd died in the running accident. I don't know if it was the buildup from all those months in the graveyards, or the disappointment of what I thought the forests should have been on my birthday, or just the experience of finally seeing a forest with small yellow-cedar saplings and towering elders, but in that moment, I understood his reverence. The canopy above reminded me of a chapel or shrine with a stained-glass ceiling—shades of green, stems slicing across the verdure like the lines of a mosaic, white sky peeping through.

I stood in the middle of the plot by our center stake and placed my hand on the bark of a cedar. *Fifty-two, maybe fifty-four.* I'd recorded so many tree diameters that I could estimate within a few centimeters of what P-Fisch reported. I took my hand away and raised it to my nose, the sweet smell from that touch still there. Maddog and I got so excited about seeing a small, tender sapling that we dropped to the forest floor to sit beside it for a moment. I gently tapped its top with my fingertips, and the little cypress limbs fluttered like wings.

"They're so soft," Maddog said, stroking the foliage.

"If I were a deer, I think this would be my first pick in the forest, too," I replied.

"For sure."

P-Fisch said working in the forests with live and healthy yellow-cedar trees felt like vacation in comparison to the work on Chichagof. They were open down low and full up top, free from the chaos of a community in transition. With fewer measurements to make, getting a plot a day was no trouble. Only the cold challenged us. We couldn't see the Brady Glacier, but we could feel it. An ice cube the size of Rhode Island was just on the other side of the inlet.

"It has literally been raining for forty-eight hours straight," Maddog wrote in her journal on day four. "I'm starting to remember what it feels like to be in a constant state of discomfort." Days later, she wrote, "If today was anything it was cold. I mean really cold. Like I am so uncomfortable I could curl over and die. . . . All I could concentrate on was how cold I felt. Must get better at distracting myself."

I started wearing my lifejacket in the forests over my rubber and Gore-Tex; its thick foam was the only added layer I could think of to keep my core warm. I'd strap the jacket onto my backpack on the hike in, then put it on once we made it to "the office" for the day's endeavor. I'm sure I looked ridiculous wearing bright orange fisherman pants and carrying a clipboard, bounding from tree to tree in a flotation device. But it worked, kind of.

"If a tsunami hits, I'm totally ready," I said, buckling up on the coldest days.

"The trees look bigger, healthier, and happier"; "hardly any snags compared to last summer," Maddog noted in her journal. She wrote every night in the nest. Instead of talking about what might come, we went to sleep most evenings discussing Chichagof and all the terrain we'd covered. The forests in the park gave us a glimpse into what those graveyards had once been.

We never actually saw the territorial bear. The biggest surprise inside the park, instead, was a patch of live yellow-cedar trees so dense we had to push our way through. I got what I needed—ten seemingly healthy "controls" for the chronosequence.

On the morning of our final pickup, I put on the wetsuit Captain Zach had left with us for another layer. In a tribute to the last inlet we surveyed, I swam across Dick's Arm. The water was so cold I couldn't put my head under, and about halfway across, I thought my heart might stop. When my hands touched shore, I stumbled to my feet and jogged along the sand to get warm, my crew cheering from the other side. When I made it back, Tomas and P-Fisch had coffee waiting.

"When it ends, the curtain drops," Tomas said. "We gave these forests our all, and that's that." It was an awkward and insufficient parting back in Gustavus to mark the completion of fifty plots over those two summers.

We had kept each other alive out there.

THE TEMPERATURE DEVICES I installed later revealed that when the snow came for winter in the park, it stayed. The blanket of insulation settled, and temperatures seldom fluctuated between freeze and thaw. Spring came later there than it did on Chichagof. What chilled our bones in Act II was also what was keeping the trees alive, but my results later showed that future warming could put them at risk too.

In a plant community experiencing the death of yellow-cedar trees, traits that define how species interact with each other and with the surrounding environment shape, in part, which plants thrive, and when and where.[2] I had that piece of the puzzle solved with the thousands of measurements we'd collected. Traits, for example, help determine which plants will reach for the light, and which will wilt, unable to tolerate the dramatically different conditions.

Long before humans really started messing with rates of change, Charles Darwin used the term "adaptation" to describe how an organism evolves to become better suited to its habitat.[3] It was 1859 when he published *On the Origin of Species*, putting forward, with evidence from his journey around the world, the theory that populations evolve over generations through natural selection. But when it came to people adapting to climate change, I wasn't thinking about adaptation as an evolutionary process over millennia anymore. I was wondering how people decide what we can do now, today, and tomorrow.

What were the traits that could lead a *person* to thrive in a rapidly changing world? Maybe they weren't traits in the biological sense of the word, but characteristics or conditions that varied among people.

Whatever the case, that's what I wanted to know when I returned to town from the coast.

From what I'd observed in the forests already, it appeared as if western hemlock trees took over after the death of yellow-cedars. Mosses and ferns became less prevalent, and shrubs flourished. I thought if I included people as part of the ecological "system," as part of nature, it was possible some people might have found ways to make the best of the shrubs and western hemlock, depending on how they used and valued the forests. Others might only experience loss. I wondered if the increase in some understory plants could create more forage for deer. To adjust to the unexpected environment emerging, I figured people would need to see not only the negative impacts, but also the strategies some were using to try to maintain balance in the forest.

I'd spent a large chunk of that year in California, reading every scientific paper I could find on what leads an individual to take action in the face of change.

"Mind the Gap," a paper published by a graduate student named Anja Kollmuss and her professor Julian Agyeman in 2002, was one of the most influential scholarly reviews of work pertinent to the growing fields of environmental education and behavioral change. It explained the troubling chasm between knowledge of environmental problems and what researchers call "pro-environmental behavior"—actions like recycling to address waste problems, or reducing water use to combat drought, and it formed the basis of my work to come.[4]

The gap, I learned, had puzzled psychologists and behavioral scientists for decades. They wondered, in light of the increasing awareness of the roles people play in creating environmental problems, why we still fail to act. In cases where individuals have altered their behaviors—choosing, for example, to reduce plastic waste—they wanted to figure out why. Researchers in the early 1970s thought if they could uncover what that was—one motivating factor or a cocktail of many—it could, potentially, help solve environmental problems, such as plastic waste or water pollution.

I thought the same might be true for climate change, where individual and collective actions can contribute to mitigation and adaptation—not

Darwin's adaptation, but what climate change experts call the process of adjustment to actual or expected climate and its effects.[5] I didn't know what "adjustment" could look like in real life, and the scientific definition itself was still a work in progress.[6] But I believed it was inevitable that the new dynamics in the coastal forests would cascade to the people who knew them well. Maybe they would change the ways in which they related to the trees, or to each other. Maybe, if they knew what was causing it, they'd feel more compelled to do something—anything—about climate change itself, because the seemingly abstract warming was having real effects at home. That's where my hope began to reside.

Kollmuss and Agyeman's summary highlighted early assumptions about the relationship between knowledge and action: knowledge (K) could lead to certain attitudes (A) about environmental issues that would then spark a change in behavior (B). "K-A-B," researchers called it. But in the decades that followed, the simple knowledge-attitudes-behavior model failed. Raising awareness didn't do the trick. So then researchers uncovered many other contributing factors to behavioral change—such as whether people learned about an environmental impact indirectly (through school or media, for example) or by experiencing it directly; whether they felt concern; whether the issue was one they felt they could address; and whether they'd developed some level of attachment to the place impacted.[7] After all, we protect what we love.

If Alaskans who used and valued yellow-cedar trees had found ways to forge ahead amidst the trees' decline, I believed their secret was hidden somewhere in the complex web of the K-A-B model and the many other factors Kollmuss and Agyeman had synthesized. I would try to understand what the individuals I interviewed knew about the dying trees and what was killing them. I'd ask questions to understand their attitudes. Were they concerned about the trees' death and its cause? Did they care? Why? I'd explore how these Alaskans used the forests and related to them, and how those dynamics might be shifting. Were they taking new actions? Had their use of the forests changed? How? Were they coping? Why? In the hours of interviews I'd ultimately collect, I'd search for patterns in the answers and test relationships between knowledge and action.

Only later would I learn that what I'd experienced over the course of fifty plots and two years—the fear, the vulnerability, the wondering what I could do, the reverence, the loss, the searching for the positive amidst the negative consequences—all of it could have made me another interview subject in my own study.

I FOUND GREG in early September at the chopping block where we'd sat two years prior, me on the bucket, Greg on the stump. We walked down the gravel path to his home. On the porch outside, his wife, Judy, passed him a tray of canning jars to bring inside. The scheduled interview was my fifth pilot since returning from the outer coast and the last one before I left again for California. Three months had passed in the archipelago since that pasta dinner, and it would be another six or seven until I returned again. He nodded toward the door, so I opened it, stepping into the warmth he'd cultivated over fifty years rooted in place. The floors were painted plywood. The walls were rough-cut wood. There

Vaccinium alaskaense (Alaska blueberry).

was a kitchen table by a big window in the corner, cozy couches, and books neatly aligned. Steam rose steadily from a large pot on the stove. The tick-tick-tick of what sounded like a metronome, pulsing at the pace of a human heart, filled the room.

"It's for the electric fence, for the porcupines," Greg explained, gesturing to the garden.

"What do you mean?"

"I've put so much work into that garden, and we need it. I can't risk any tiny chance of one of those porcupines getting in. So I moved the fence box inside and put a ticker on it, and now I know it's running all the time." He pointed to the small metal box that was emitting the tick-tick-tick. It was mounted to the wall beside the window that looked out over the green rows.

"They're standing out there ready to pounce on our carrots at any moment."

I laughed. "Well, I would be, too," I said, scanning the vegetables.

"If I lose the electric charge, the tick stops. Then I know I've got to get out there to defend our crop from all the other animals interested."

He lives with this constant ticking, 24 hours a day?! I thought. *That would make me crazy!*

"It's a very real threat," Greg said. "The potatoes make up a good part of our diet."

Clearly, turning the metronome off for our interview wasn't an option, so I tried to relax into the reminder of time, the tick-tick clicking like one heartbeat after another. There was something soothing about it.

We sat down on two couches across from one another and Greg asked me questions about the outer coast. I glossed over the forests, hesitant to share, just yet, what I'd seen on Chichagof and inside the park. I was there to ask about his knowledge and his attitudes about the changing forests, and his relationship to them. All that had to happen without my own experience or information about the dieback influencing anything. The same would be true for all the interviews that followed, for all the different perspectives I would gather; I would need to keep the findings from two summers among the trees hidden.

I wasn't doing the pollster approach of forcing people into predetermined answers, such as "Yes, I've seen patches of dead yellow-cedar trees," or "No, I haven't," or "I don't know." Like the strength of big data in revealing climate trends, surveys of hundreds or thousands of people have a sexy appeal. If conducted with a random sample of a given population—say, Sitka residents, or Forest Service employees—they can reveal powerful results in rigorous statistics. But the kind of information they elicit is limited to the boxes or scales the researcher creates. Instead, I took the approach of "semi-structured" interviews to create an open space for discovering the unexpected. The analysis would be challenging and incredibly time-consuming. The patterns would reveal themselves only by what would emerge through the depth of conversation with a relatively small number of people. I had a list of must-ask questions to guide what I hoped would be a comfortable dialogue based upon trust.

"So, question for you," I said, shifting the direction of conversation. "Do you distinguish a forest with yellow-cedar differently? In what ways?"

I strived to ask every question calmly and confidently, leaving little space for him to turn them back on me. A more natural style of back-and-forth conversation would come for us in the years to follow.

"Well, yellow-cedars are one of my two or three favorite organisms in the whole world," Greg replied. "And in part because they personify a lifestyle which we humans only instinctively put at risk—" Something caught his eye outside the window.

"Oh, here comes that big brown dog again."

"You could put him on porcupine detail," I said, pointing toward the garden. Judy stepped outside to tie up the wandering pup while she found the owner. I stopped the recording and restarted it when Greg's attention returned.

"One of the things that humanity seems to do, almost instinctively," he continued, "is to substitute youth for antiquity in natural systems." He spoke with careful pauses, rolling out words like a poet searching for the right prose. "Nothing is ever allowed to get old around people. And cedars are one of the quintessential things that are disadvantaged by that. So, they appeal to me from that standpoint, as kind of underdogs. They also appeal to me because their presence in the forest is so lovely, the way the wind sounds in them. The color of their bark. The way that each tree is so idiosyncratic. And as an object of scientific inquiry, it's such a puzzlement why they are distributed how they are, and how they have been through time. It seems they almost make it hard for you to study them because their pollen is so poor in the record. They are very mysterious creatures."

Greg said that he put yellow-cedar trees in the same category as marbled murrelets or elephants—"K-selected species that grow slowly, live a long time, and reproduce sparingly." He was referencing a popular theory in ecology—r/K selection, first introduced by biologists Robert MacArthur and E. O. Wilson in 1967. I'd learned about it at Stanford, but never considered it as a way to describe the effect people had on other species over time. MacArthur and Wilson's book, *The Theory of Island Biogeography*, used principles of population ecology and genetics to predict the number of species that would exist, theoretically, on a newly created island. In their mathematical equations, the size of the island and its distance from other lands created a balance between the species that immigrated and the species that went extinct. But at the core of their calculations were two variables—*r*, referring to rate, and *K*, carrying capacity—that still influence ideas on survival today. Whereas r-selected species, like rabbits, shrews, or grasses, reproduce rapidly in large numbers, K-selected species, like a hippopotamus or a giant sequoia, invest more into relatively few offspring. In Greg's eyes, people have been able to manage the r-selected species relatively well, but we drive Ks—like the cypress—toward extinction.

"There are these great ancient organisms that are just being whacked down and sent in the round to Japan," he continued. Through the r/K-selected lens, harvesting ivory from an elephant in Africa carries the same weight as cutting a yellow-cedar tree in Alaska, or killing one indirectly through climate change. I made notes to look more into yellow-cedar exports and inquired further on K and r—the elephant and the shrew.

"One of the reasons that I've become a bit disenchanted with science these days," he explained, "is that it turns into increasingly elegant ways of increasing the pressing. You just move your fingers a little differently so it doesn't quite press there right now, but we'll press over here a little harder. But we're not going to quit pressing. I think we have to manage from a different part of ourselves, actually—a part of ourselves that can see some point in restraint. Science is not going to give us that. That's the conclusion I've come to." He didn't seem to be talking about "we" as scientists anymore, but rather "we" as humans, or humanity itself. "As far as K-selected species go," he added, "what do you want to talk about there?"

"I think you've answered it," I said, still trying to connect the dots. *He's saying that we've pushed ourselves into a world where the old and wise cannot survive; we've come to value the long-lived more for their immediate use than for their continued existence. There's something we need to resolve for ourselves to fix this. What?* He wasn't just talking about trees. I waited to circle back and proceeded with what I thought would be a few less probing questions on the list.

I asked a series of questions to explore any differences between how he used, or related to, a forest with live yellow-cedar trees as opposed to one affected by dieback. "Are there areas that you have seen or spent time in where the decline is occurring?" I asked. Then, "So the places where you've spent time with live or healthy yellow-cedars, those are mostly up around here, around Glacier Bay?" And, "Are you mostly doing research there? Or going out to enjoy, or for other reasons?" His answers were straightforward, taking me through his own research in the park, to magnificent trees, and then to boat rides out to Sitka—Peril Strait—where dead trees covered the hillsides.

When I asked what the decline meant to him, he gazed across the room for a moment at a wall of photographs—family members, I presumed, generations of Alaskans on ridgetops, in fields, and by rivers.

"Well, the cedar decline affects me the same way the decline of elephants affects me," he answered. "I guess I would harken back to the generic thing I said to you about human beings making it difficult for antiquity to exist. When nature conspires on that, it's a double whammy. So, I foresee a time when cedars will be basically scarce in Southeast Alaska, and that affects me the same way that I think about murrelets being someday scarce in Southeast Alaska. That's a philosophical thing. That's an emotional thing. I don't like to think about ancient things being gone."

Greg said that his relationship with the forests had gone from "predominantly one of objective interest to a friend," and that his empathy "for the creatures of the forest" had deepened. Up until only a few years ago, he would have called himself a scientist, but his attention had shifted when the "non-science parts" of himself started atrophying. Years ago, Greg was interested in a forest's nutrient load, but he'd become more engaged with the forest as an object of veneration instead of a pathway for dissecting a research question.

"One of the reasons I've gotten back into geology so much," he said, "is that if there's one thing that interests me most deeply, it's the concept of time. There are actual forests. There are potential forests. There are forests I can imagine in my mind's eye . . . because there's stuff in the rocks that are Middle Tertiary. So there are sixty-million-year-old forests here, and there are forests coming to be. We're sitting on one."

He saw a forest the same way he saw a human identity.

"A forest is a concept. A forest is an actuating algorithm that we are catching at a moment. But the beauty, to me, one of the principal beauties is to try to imagine the stream of matter and energy through this moment from where it's coming from to where it's going. So that's the forest."

Greg's description of a healthy forest was far from anything I expected from a scientist, but I could relate, and that gave me some solace.

"A healthy forest," he said, "is a lot like a healthy human organism. It's one that, first of all, it's an emotional category to me. It's one that when I immerse myself in it, the feelings that it imparts to me are ones of calmness and dignity." I leaned forward.

"Yes, I'll use those two words," he affirmed. "When I go in the woods and I feel anything other than that, I tend to define it as less healthy. It's not just the trees, it's the condition of the soil and the critters. It's something that I know should be there, but isn't. It's a category that doesn't have much scientific veracity in the way I look at it. It's a feeling."

There was space in my interview protocol to follow threads that emerged in addition to the firm set of questions I would ask everyone. And so I deviated.

"Well, when you started the conversation," I noted, "you said that the perspective you have gives you a sense of forests coming and going. You can be in a place and know there was a forest there thousands of years ago and know there will be one in the future. But yet I hear a lot of loss when you talk about yellow-cedar decline. Are there certain hopes or other opportunities when you think about what the future holds for these forests?"

"Well, this probably gets more philosophical than you would like. But I don't do hope," he said. "One of the reasons I think geology has become important to me is that it helps me pass the pain I just mentioned to you. I'm getting better at visualizing deep time in both directions. It makes me realize that the present moment of human depredation is definitely going to be fleeting. Other things will change in ways that I can't imagine. But there will be ancient things again in the world at some point, and there have been. So it gives my spirit respite to live in remote times, either future or past."

"You don't do hope." I was shocked. "But yet you can see across time? What's that about?"

"You really want to get into this philosophical stuff, huh? You can't stand not getting into this philosophical stuff. That's one reason I like you, Lauren, but I pity you." He grinned.

"You're not going to tell me the hemlocks are regenerating or the forest will be okay?" I pushed back, playing with the notion that yellow-cedar trees dying and turning over to western hemlock was no big deal to anyone.

"No. What I mean is that—well, here's the substance of it: In the modern world, I think it's intellectually dishonest to be hopeful, but it's equally stupid to be hopeless. You can't live out of a hopeless life."

I'm totally confused. He's saying there's no hope?

"So where does that leave you?" he asked me.

Good question. Hopeless? No, neither hopeful nor hopeless.

I said nothing.

Greg continued, "Well, what occurred to me a few years ago was that I don't have to get caught in that trap. The best thing for me to do is to develop my inner voice and to steer as close to that as I can and to act as if what I do matters. And allow the future to decide what comes of who I am. It's totally silly, this, these various little ra-ra books about a thousand points of light and all that stuff. Every trend in the world that I care about is not only going in the wrong direction, but it's accelerating in the wrong direction. As long as the hand gets heavier, cedar trees are in trouble. Murrelets are in trouble. And eventually we're in trouble."

So there isn't anything we can do?

"There was a fairly brief period in my life when I was pretty well philosophically prostrated by this because I couldn't bring myself to play these little hope games and say, 'Oh, see that little thing over there, notice now that the car is using a few gallons per hour less,' or, 'Look, someone just put a solar panel on their roof. And so things are getting better!' Well, they're not getting better. I didn't want to play that game with myself, and yet I didn't want to be trapped in the abysm of being depressed over it. I want to live a more joyful life than that."

So do I. So do I.

Maybe it's not about hope or hopelessness, but faith instead. Living in longing for some other future condition sounded painful to me; helping to create one—playing a part of a more positive, less grim tomorrow—was far more inspiring.

In terms of K-A-B—Greg's knowledge about the dieback, his attitudes, and his behaviors—our afternoon together yielded the more concrete, less philosophical, information I needed. What he knew about the dieback and what was causing it, whether he was concerned or apathetic, and how he was (or wasn't) changing any behaviors around the affected forests—those were all straightforward. But he was just one person, with a perspective that made me question my own.

Somewhere in the process of becoming a scientist, a student or someone new to the field often learns to focus only on the metrics of success that the system says matter—testing hypotheses, publishing papers, uncovering one new truth only to pursue another. I saw that early on at Stanford. At my stage, just beginning in science, you learn what the system rewards and what it doesn't, that amidst facts and figures, there's no place for emotion. Then at the end, coming out, like Greg, maybe you know what life rewards and what it doesn't. An ecologist who works in a place, in a community, not just in an office or in front of a computer with its 1s and 0s, comes to know that place like a friend; watching it change, or its members die, is a challenging loss, like any other in life. Still, Greg was just one person. Finding patterns and deciphering answers for how to cope and how to adapt would take dozens more interviews.

Honestly, I left feeling like Greg had let me (the concerned citizen) down, that his solution was to give up. That the sacred cypress was actually *his* muse, and its death, like the canary in the coal mine, meant game over. That climate change would ultimately have the same effect on humanity. That it is better to accept what we cannot change, to live the best we can amidst all of it, instead of fighting for something else. Giving up didn't sit well with me. I didn't think living joyfully and pushing for positive change were mutually exclusive.

At the same time, Greg had driven the hook of the scientist even deeper for me. If hope, as the poet-philosopher Derrick Jensen says, is a useless emotion, where could faith reside?[8] In our knowledge? In our attitudes? In our behaviors? Was there something hidden behind Greg's complicated answers and the many others I'd record that could possibly enable me—another K-selected creature in the mess—to stand tall and thrive?

Western hemlock growing beside a dead yellow-cedar.

CHAPTER 7

Coveted

It was a cold, damp Monday night in January when my brother called from New York. I was crunching plant data from the outer coast in my long underwear, my toes resting at the base of a small space heater pressed up against my desk. The window in my bedroom rattled in the wind. Through the thin glass, I could hear the waves crashing outside.

After I left Greg and Alaska again, I returned home to Palo Alto. I had wanted to stay, to linger longer in Juneau or Gustavus, but I couldn't. All the work that lay ahead required me to be near campus, where I could troubleshoot problems and analyze data with colleagues and lab groups. Maybe the seasonal clash of worlds between quiet Alaska and the bustling Silicon Valley was becoming too much, or maybe my Palo Alto home never really felt like home at all. But when a friend called me about a room in a big house by the sea with a community of people in Santa Cruz, I packed my things and moved to the coast.

I'd only been there for a few weeks before that phone call. Before everything changed again. Before my brother told me that my father had died while taking a nap that afternoon. His neighbor had found him at home on the sofa in his living room—there, asleep forever.

I got up from my desk and stumbled down the winding wooden stairs in my socks, flung open the door, and ran into the darkness toward the roaring sea. I ran through the streets in my socks until I hit the cliff, and then I just kept running along the ocean overlook, holding the phone to my ear, listening to my brother weep while I gasped for air.

The bottoms of my socks got so wet that their weight began to pull them off. My pace slowed as I tripped over their soggy tips that kept flapping against the cement. Finally, when a man with a sleeping bag under his arm stepped into the street from the shadow of a tree, I stopped. The tree was a Monterey cypress, another California relative of my northern friend, and it was twisted and gnarled from the salty winds. The man just stood there underneath a streetlamp and stared back at me, this woman out of breath, alone in the night, in long underwear and socks.

I traced the shadow of the trunk back to the tree and looked up toward its branches. They fanned out in layers with their flat tops reaching for the sea. The man took a step forward.

What am I doing? Go home.

"Are you still there?" my brother asked.

I turned around and began running in the other direction—or what I thought was the other direction—until I arrived at a white picket fence, a familiar fence that bordered a yard a couple blocks from home. I walked the last stretch and stood before my front door. It was still open, and warm orange light from inside poured onto the foggy street. I don't think we'd been on the phone for more than a couple of minutes, but it felt like I'd been running in circles for hours.

"Lauren?"

"Yeah, I'm here," I said. "Does Mom know?"

She knew. My brother's wife had been the one to tell him. My mother had decided she couldn't be the one to tell her children. She had never wanted to bear this moment, or the one my brother and his wife had shared in the car outside the airport in New York that evening. She'd never wanted to know what either one of us would say or do when we found out we'd lost our father.

I called a few friends for help and no one answered. I waited and then called my best friend in Seattle back twice. Still, she didn't pick up.

No one answers the phone anymore. Am I really going to text this?

Yes. I am.

I did, and that worked.

My friend stayed with me on the phone that night until I fell asleep. In the morning, I had trouble deciding which black dress to pack, but it seemed like I should bring one.

Within hours of landing off a red-eye flight, I was sorting out what kind of autopsy we wanted and what food we should serve at the memorial while trying to process emotions and think about what I'd want to say that day. Or not say. My father had never wanted a sad service. He had wanted a party.

Just as most women gloss over what really happens in childbirth, there are things no one ever tells you about a parent's death. Someone has to confirm the body. Someone has to decide what coffin or urn to use or what photograph to put on the handout that someone has to design. All those someones were me or my brother, Ryan, or maybe his wife, Mika. There's a cry that is like no other. It is my mother's when my father dies, even if they were divorced. I imagined myself sorting through his things in his home, smelling his clothes, and reading letters, deciding what to keep and what to give away or throw away or sell, what to photograph, what to remember, and what to forget. The autopsy revealed he'd had a heart attack. Someone has to tell everyone, and someone has to decide who everyone is.

"Wait, what time is the service?" "What color should the flowers be?" "Shoot, we'll need chairs." "How many?" "Who knows?"

I don't know.

Amidst it all, I kept trying to remember everything he'd said in our last conversation, and then the little stuff—the way he danced offbeat to the music while making pancakes on a Sunday morning, how he sipped coffee from the same mug every day while reading the *New York Times*.

"You're thirty-one," he had said to me on the phone a few days prior. "I was thinking the other day, you've lived almost half my life now. It's going to fly by despite all you do to slow it down. Make sure you love enough and share enough with someone. And don't spend too much time in front of a computer."

I knew I'd have to let go, and I somehow trusted that process—however it would unfold or whatever it would become. But I still wanted to hold on to what I could before my memories slipped away with him.

THERE WERE SO many logistics to manage; it wasn't until I returned to Santa Cruz a week later that I even began to realize he was really gone. I came home with my own suitcase plus my father's, which I had filled with only a few things I wanted to take: The soundtrack to my childhood in a collection of Bob Dylan and Beatles records. A V-neck cashmere sweater that was too big for me but still smelled like him. Three thick, hardbound biographies I'd found on the stand beside the couch where he died—Steve Jobs, Bob Dylan, Keith Richards. When we had arrived in his home, the books were neatly stacked, his glasses resting on top. I could see my father taking off his spectacles to rest his eyes one last time. I stacked the books in the same order on my own desk by the sea.

My father loved reading biographies. He loved them for the wholeness of character they presented, for the common threads of humanity, for the portrait of an individual as a sum of parts—the good and the bad. He was a very smart man. He struggled with the highs and lows of manic depression, and his presence in my own life often mirrored that trajectory—fully there when he could be, and then unpredictably not. Repeat cycle, again. I think he read biographies to see the struggle in others—even in great minds and legendary leaders. At his memorial, one of my father's best friends from Harvard had told me, "Your relationship determines everything now. If there's a gaping hole, you'll grieve that pure loss. If he'd never been there at all, you'd probably feel a sense of closure. If he was anything in between, that's more complicated. You have to sort out whatever you lost from whatever just got simpler."

Everyone at Stanford told me to take as much time as I needed. "I tend to process tragedy well if I just put my energy there for a while,"

I wrote my adviser, Eric Lambin, when I returned, but I felt I couldn't put life or death—those dying trees—on pause.

"If you think you can do it, you should prepare and go," Eric told me in February. A Belgian professor of earth sciences and a member of the National Academy of Sciences, Eric had spent his career building a scientific field called "land change science" to understand what drives changes in land use around the world. He consistently challenged me to grow as a scientist, respected my intuition, and inspired me to hover productively on the edge of discomfort. Emotions aside, if I didn't make it back to Alaska in the spring for interviews, I'd have to wait until fall, when the seasonal pulse of summer activities would settle again in the North. Practically speaking, waiting six months for interviews would probably mean waiting another year for answers, and neither my PhD nor my patience could handle that. The more I'd learned, the more pressing time felt. I wanted to immerse myself in other people's stories. I desperately needed to keep moving forward, and so we agreed I would.

I thought of the decision as putting my father's death on hold for the trees. In reality, it meant I would tap into a larger grief. I was signing up for months of conversations with strangers who were experiencing change and loss all around them. It would be impossible to forget my own loss—the loss of my father—amidst the growing knowledge of an uncertain future for the yellow-cedar. My data conclusively showed that once climate change hit a patch of yellow-cedar trees, the chance of finding a sapling—the next generation—went down. The graph I'd constructed for that probability looked like the downward-sloping trough below the crest of a wave. I posted it on my wall over the stack of my father's biographies.

"Yellow-cedar appear maladapted to forests affected by decline for the foreseeable future," I wrote in the draft of a study I'd later publish in a journal called *Ecosphere*.[1] Greg Streveler would have shaken his head and called my work the early signs of monitoring a species to the point of extinction.

Before we had parted a year earlier, Greg had told me, "Grace is what we decide to take with us and what we leave behind." He said the idea was from the philosopher Erik Erikson. I went looking for the original source after my father died, after I made that graph, after I had listened and analyzed the transcripts of my pilot interviews. I couldn't track down Erikson's exact words. I wrote Greg to inquire further—but to no avail. It didn't really matter. What mattered to me in my research was the notion that we have some choice in how we move forward.

Loss was a common theme in those interviews, and with a sense of loss ever present in my life, I could relate and see the pattern more clearly. "Grief over the loss of cedars is a powerful emotion and unfortunately it's a rare emotion," one thoughtful man told me, explaining his experience witnessing the death of yellow-cedar trees. "You have to cultivate a certain level of affection to understand the significance of the loss. Just to love a cedar is a meditation unto itself, and then to grieve for the loss of a cedar is another piece of the meditation. Then to act from that meditation is yet another piece of it."

If continued warming is inevitable (at least to some extent), we need to focus on adapting to what we cannot change. Our choice is the decision to make the best of conditions we never imagined, never wanted, and maybe even feared. Psychologists and behavioral scientists focused on K-A-B as the determinants of our response, trying to unpack the ways in which our knowledge and our attitudes shape our behaviors. Darwin focused on the biological aspects: some individuals fare better than others in changing conditions, and over time, species evolve. Greg called the path forward grace, and that perspective was more empowering to me. In a difficult situation, grace meant my actions and my outlook—my choices—matter.

"I'll hold faith for you that there is nourishment there," a friend wrote me soon after my father's death, "to be found in the bits and parts through the years in the woods, on the beach, in the laughter of your own son." I didn't have a son (or a daughter), and he knew that, but the idea of a new life remembering the old was comforting. "Human death is so sudden, so absolutely irreversible, that it will come as a surprise when you find, all those months or years later, that your father is not totally gone."

I flew north in April with approval from the university to go forward with my interviews—to chase after patterns in the K-A-B model and anything else that emerged. I was hoping to find a process that enabled someone who knew the yellow-cedar trees were dying to change their uses of them, or the forests themselves; ideally, it would reveal some sort of prescription that enabled someone who blamed climate change not to feel hopeless and helpless but to take action instead. I wanted to uncover a way to confront loss and create a new path forward within the emerging environment.

KEITH RUSH WAS sitting at a round table in his office when I arrived in April 2013 for the first of my Juneau interviews. He was dressed in typical forester wardrobe: heavy green canvas pants with big side pockets, a fleece pulled over a t-shirt, and a baseball cap. One of the scientists who had worked with John Caouette, the forest statistician, had suggested Keith as a "study participant" for me, although I never used those words in my inquiries.

I'd started rolling the snowball from California by calling everyone I knew in the cedar circle, asking for recommendations for interviewees. I was looking for people who used and valued the forests in various ways. "I'm interested in talking with you about yellow-cedar forests and learning more from your knowledge and experience in this region," was my common introduction. I didn't like the idea of putting anyone into a predetermined box, but in order to ensure I interviewed people who represented a range of relationships to the forests, I needed a few. I used categories to group people in my search by uses: customary and traditional uses, including Alaska Natives engaged in hunting, fishing, and gathering activities; sport hunting and subsistence activities among non-Natives, such as berry-picking; recreational and tourist uses; conservationism; uses by scientists and naturalists; and forest management activities. I'd reviewed regional management documents from the Forest Service, from local news media, and from the websites of organizations, agencies, and companies involved in forest issues to determine the categories I would fill.

When I went to interview Keith, the conservation forester for The Nature Conservancy, I had already put him into the "conservation" box in my spreadsheet. In his office, pictures of backcountry ski trips covered a corkboard on the wall. Originally from the Midwest, Keith had a slight drawl.

My questions began with simple demographics to allow the people I interviewed to get more comfortable.

"How long have you lived in Alaska?"

"Twelve years."

"What first attracted you to move to Alaska?"

"The wildness," he said. His wife was a wildlife biologist. "We used to be outside even in our work, and our free time is always outside. There's no better place than Alaska to find the wild."

"In terms of your relationship to the forests here, which one of these groups would you most identify with?" I asked. I always checked whether the box that I'd assigned before we met aligned with the one each interviewee picked.

"Probably conservation," he said, but he had "a smattering" and "experience" with most of the categories.

What began with quick background questions then became more complex. "What kinds of activities and natural disturbances do you feel have altered yellow-cedar forests in Southeast Alaska?" I asked.

"What the researchers tell me, which I accept until any better information comes along, is that it's climate change that has led to the decline."

Keith described various locations where the decline was occurring and explained the intricacies of the causes. He called it a "perfect storm" of two factors for the "low-elevation yellow-cedar tree"—climate had already been warming after the Little Ice Age, and human activities were now exacerbating that warming by contributing more greenhouse gases to the atmosphere.

Out hiking, he'd observed cedars at higher elevations and thought they were healthier because of that fact. "Those sites retained snow," he said. "Cooler temperatures and snow through the spring, so they weren't being damaged by the early spring frost."

"You touched upon it a little bit," I said, "but what does a healthy yellow-cedar forest mean to you?"

"Well, a healthy one, to me, in my own mind, is gonna have a lot of mortality in it as well," he replied, bringing me back to the reverse-J curve. Someone trained as a forester, like Keith, learns the ecological value of the deceased trees—the habitat they create for birds, the nutrients they add to the forest floor as they decay. In any forest, many trees simply can't survive to reach enormous sizes. They die from injury or disease, or they get outcompeted by more vigorous individuals. Those "normal" dynamics of life and death aren't about climate change.

"I think the die-off, although man is contributing to it, it's a natural process, and so I don't get really upset about it. I just take it for what it is."

He reached back in time to draw parallels to the past when people had survived rapid change.

"When the Little Ice Age was just beginning in Greenland and Iceland, people were seeing spectacular events in front of their eyes. We're just seeing the backend of that, going back to what they experienced beforehand." I found his confusing explanation fitting for the complexity of climatic events.

There was some truth to his arguments about natural cycles and healthy forests. The Little Ice Age was a period from about 1200 to 1900 CE when many glaciers, in Alaska, Greenland, and other parts of the world, extended to their farthest point since the end of the Pleistocene.[2] Hoonah, the largely Native community where I was headed next, had been settled when the advancing ice had pushed the Tlingit people out of Glacier Bay in the mid-1700s, forcing them to find refuge in a rapidly changing climate. Paul Hennon's research showed that the onset of the dieback coincided with the end of the Little Ice Age; as climate warmed and glaciers retreated, the snow, too, slipped away in places.[3] So, in Keith's eyes, there was an environmental change already occurring (the warming that followed the Little Ice Age). Now people were just intensifying the trend—thus, the perfect storm for cedar.

"Anytime I'm on a walk recreationally or for work, I'm always pointing out big logs laying on the ground and saying, 'Oh, wow. Look at that down dead wood.' Some people think I'm crazy for it, but it's because I know how valuable they are to so many functions and species."

What caught my attention most, however, was the fact that he wasn't grappling with the loss of yellow-cedar. The death of the tree didn't appear to impact him directly. If Greg was on one end of the spectrum—feeling deeply the loss of a species he loved—Keith was on the other.

"Do you think of yourself and yellow-cedar forests as connected in any ways?" I asked, trying to understand why.

"Probably not," he replied. "Because I got acquainted with a yellow-cedar forest at the end of my forestry career, I feel like a visitor. I don't feel that connection. If I would have grown up here, spent my whole career here, I would have more of a connection, but just because I'm a newbie, that doesn't mean I don't care. It just means I don't have the connection I probably would have if I would have spent thirty years plus out in those forests."

Whereas Greg or John Caouette, and many others I'd later meet, described the yellow-cedar as majestic, and prized the species as especially important, to Keith it was another tree in the forest.

Attachment, I thought, sitting there listening to his seemingly removed explanations. *He's not really attached to this one species.*

How we cope with the loss of an individual has a lot to do with how and how much we value it. To experience loss, one has to experience love or some form of attachment. This connection can seem so obvious in the case of a friend, a family member, or a loved one—and I was in the thick of that with the death of my father—but the same holds true for attachment to nature and place. Psychologists and behavioral scientists distinguish two forms of nature attachment: functional, when a resource provides amenities that are necessary for desired activities, and emotional, when a psychological investment in a setting or a resource develops through experience over time.[4]

Keith's job focused, in part, on facilitating the "young growth transition"—a shift from logging old growth in Southeast Alaska to planning for young-growth harvests in areas that had previously been cut. His work required him to look at the forest as a whole, to see not only the economic value of a specific tree or species, but also the many ecological values of all its parts. He simply had never developed his own functional or emotional attachment to yellow-cedar trees in particular. Sure, he was worried about climate change. But what it meant for the cypress was less pressing for him personally than it was for many of the other people I would later interview, who were far more attached.

He showed me a few samples of young-growth wood used for paneling. A few years back, he'd been involved in a workshop with the Forest Service and a few mill operators to discuss whether the dead cedar could be a viable product. They'd chopped four dead trees to the ground and examined their properties.

"One thing we found out was that the local sawmill guys realized there was a lot of value in those trees," he said. "The sapwood started to deteriorate and fall off when it was medium old, and when it was old, all you were really looking at was the original heartwood. The snag was actually smaller in diameter at that point in time than it had been, but that heartwood was still sound."

If harvesting old growth held a high ecological risk, Keith said, the risk of harvesting dead cedar was relatively low. He thought there was potential to trade the uses of dead trees for live trees, as a means of preserving the survivors.

"Maybe if you could get the focus or intention toward the dead snags that still have really good value," he said, "then you could take some of the pressure off the healthy ones and let them regenerate. Keep a good seed source out there."

My interview with Keith was so drastically different from the one with Greg that I walked into every subsequent interview wondering which way it would go. I tried to understand each person's perspective, and not to let anyone else's understanding, or my own, influence each new conversation. It was like standing in a forest listening to birdsong at dawn, trying to pick out each voice and hear what it had to say. Only after focusing on each one individually, for interview after interview, would I listen to the concert the birds were creating together.

Fifteen interviews later, I boarded the ferry in Juneau to travel to Hoonah. I carried a backpack stuffed with my sleeping bag, a rain jacket, and a change of clothes, along with notes and recording equipment. I brought a mountain bike to get around town. I had arranged to meet a logger named Wes Tyler there, and I had the names of a few other people to track down.

THE PEOPLE IN Hoonah, *Xunaa* or *Xuniyaa* in Tlingit, say its name means "lee of the north wind"—protected from the north wind—and it was.[5] What I knew about the town at that time was limited to the history of its settlement during the Little Ice Age and its extensive logging more recently, in the latter part of the twentieth century. As the ship rounded the point between Icy Strait and Port Frederick, it broke through the fog into the sunny inlet where the Tlingit had settled centuries ago. Windblown waves subsided in shelter. I could see scars from clear-cut harvests on the hillsides—tracks of land now regenerating young-growth forests after everything was stripped years ago.

There aren't many places for a visitor to stay in Hoonah, but, given my research partnership with Paul Hennon on the outer coast, the Forest Service had agreed to let me stay in the bunkhouse. I rode my bike along the harbor and into town, which was just another intersection of a few roads with a tiny shop for groceries. I leaned my bike on a hemlock tree and went inside to grab a few supplies. The shelves were filled with canned soups and sauces, bags of rice, sugar, and flour. *Slim pickings.* I picked up a box of oatmeal and a couple boxes of macaroni and cheese and continued on. At the main office for the Forest Service, I signed a few papers, agreeing to various terms of the government housing.

"You're welcome to use our conference room for any interviews you want to do while you're in town," the receptionist said.

"Thanks," I replied. "I'll make a few calls this afternoon, but that would be great for tomorrow."

"Well, you'll have the bunkhouse to yourself," she said, as she passed me the keys. "It's empty. I have you in room number seven."

A small footpath ran alongside the road, up the hill from the office building to the bunkhouse. I pushed the bike for the last stretch. My rations dangled in a plastic bag over the handlebars.

Inside the long, brown building, a sign read, "Take off your dirty shoes in the mudroom." Printouts of the government rules and regulations listed, "No firearms," "No noise between 2000 and 0600 hours," "No alcohol," "No drugs," and "No unauthorized personnel." There were a few pairs of abandoned boots in the mudroom and yellow hardhats hanging on hooks. I took off my rubber boots and left the groceries on a table by the telephone. Its buttons were worn and dirty.

I walked upstairs in my socks, carrying my pack, and strolled down the long hallway scanning for number seven. Each empty room looked the same: two twin beds on opposing sides, a pillow and a folded wool blanket resting on top of each, and a small desk at the far end, nestled between two closets.

The emptiness felt eerie.

"Hello?" I called out. No one answered. I imagined the halls filled with fieldworkers during the logging heyday, measuring tapes and rain gear strewn about. The slightly stale odor reminded me of the cabins at summer camp when I was a little girl. I dropped my pack on one of the barren beds in room seven and walked down another set of stairs. In the laundry room, I found stacks of pillowcases and sheets, each item with an ink-stamped "US." I grabbed one set for my stay. I rummaged around the kitchen cabinets, trying to find one pot small enough for my single portion of macaroni. Ladles and fry pans, stew pots, and mixing bowls—everything was jumbo-sized for the crews.

Although commercial logging in Southeast Alaska goes back to the early twentieth century, if there was a boom and bust, its beginning and ending were marked by the "50-year contracts" granted by the Forest Service to two large pulp mills in the 1950s.[6] The long-term contracts had enabled the Ketchikan Pulp Corporation and the Alaska Pulp Corporation (a Japanese company) to commit to investing in the region. The prescription for timber sale areas was clear-cutting, and yellow-cedar was seen as a nuisance species. It was often left behind where it fell. At the salt water, men chained together logs, called "boomsticks," end to end, to form a rectangular "boom" for transport to the mills. Inside the long trains of log rafts, millions of board feet of spruce and hemlock floated like toothpicks on the sea. It wasn't until the Japanese began running low on another relative of yellow-cedar, called hinoki (*Chamaecyparis obtusa*), that the market for unprocessed logs dramatically shifted.[7] As one interviewee told me about the increased demand for yellow-cedar exports in the 1970s, the trees went to Seattle before being sent on to Japan. During some years of the following decades, the dollar amount per million board feet (MBF) of yellow-cedar was more than twice that of any other species in the archipelago.[8]

The pulp mills closed before the end of their fifty-year contracts came around, but a number of operators continued to log and mill timber. By the early twenty-first century, economically speaking, yellow-cedar had become the most coveted species on the Tongass. Its presence in a timber sale comprising other species as well simply sweetened the deal. Alaskan locals wanted the wood for its rot-resistant qualities and its attractiveness, for its ability to burn hot and long as firewood, and for the jobs it would create with processing in-state. The international market wanted the wood as a substitute for hinoki and another relative in the Pacific Northwest, Port Orford cedar (*Chamaecyparis lawsoniana*).[9] Some people I interviewed said the wood had been used in temples; others had heard rumors of logs sunken in the bottom of Japanese harbors to preserve the golden wood for later use.

But the scale of logging now was nothing like what it used to be. Wes Tyler, the man I'd arranged to interview the next day, was one of the leading small-mill operators remaining in the northern reaches of the archipelago.

I made a few phone calls, cooked some pasta with powdered cheese sauce, and laid my sleeping bag on top of the government-issue sheets. Something about the now desolate space of the bunkhouse in contrast to the life that had once filled it made me think of my father, asleep on the couch with his glasses on top of the biographies. The time passing in the quiet of the bunkhouse felt like the first moments of stillness I'd experienced since his death.

I'd kept myself busy for nearly four months, moving forward in my analyses of the outer coast data, preparing for interviews, sorting out logistics again, and then interviewing one person after another. At night, I'd type up notes from the day and download the audio recordings to make backup copies. Eric, my adviser, had warned me that about ten or twenty interviews in, I'd probably start to doubt myself—whether I was asking the right questions, how I would analyze the data for answers, if I was actually capturing any trends that could tell me about how people adapt to changes in their local environment.

"At that point," he said, "just be sure you don't do anything differently. Just keep going. One interview after the next. Same questions."

I lay in bed thinking back through some of the people I'd already interviewed: the wilderness ranger with the Forest Service, the hunter working for the Alaska Department of Fish and Game, the Native carver from a town called Haines, the naturalist who'd spent years looking for the biggest trees remaining on the Tongass. I clicked back through the audio with Laurie Cooper, a guide working at a local tour company, until I found the clip that echoed in my head.

"I worry about the yellow-cedar," I listened to her say again, "because it's sort of one of those indicator species that makes you start to think, 'Are we getting up to some tipping point relative to this place being a functioning, healthy ecosystem? And is that just like the canary in the coal mine of saying, "We're starting to lose it"'? We might be able to point back, twenty or thirty years from now, and say, 'If we had paid more attention to the yellow-cedar, we would have seen the fraying of this ecosystem.'"

I locked the door to my room that night and curled up on my right side, feeling like maybe nothing I could ever do would actually help. I fell asleep listening to Bob Dylan.

WES TYLER ARRIVED on schedule at the Forest Service headquarters. He was wearing a striped-blue collared shirt with the sleeves cut off at his wrists and a puffy brown vest. It was patched with squares of silver duct tape. In worn jeans, sawdust on his shoes, he carried the smell of the outer coast forests right into the conference room with us. He couldn't have been more welcoming to me.

"Let's get down to your questions here, and then we'll go out to take a tour of the mill site," he said. "I've got my truck. I'll bring you back to town and wherever you need to be."

I'd read as much as I could about Wes on the Internet. "As residents of Hoonah," his company website read, "we are deeply invested in our community, and have always approached our business with the goal of creating beautiful value-added wood products and as many local jobs as possible from a small volume of wood." An article from the *Alaska Journal of Commerce* began, "Wes Tyler of Hoonah works in an industry that has suffered heavy losses in recent years. Despite a sharp downturn in Alaska's once-thriving timber industry that resulted in hundreds of lost jobs and mill closures, Tyler's Icy Straits Lumber and Milling Co. is surviving and recently took steps to expand."[10] He held a large presence for a small business, a testimony to all the times his name had come up in my snowball search for interviewees.

"How was your morning?" I asked, as I checked the recording equipment on the table.

"Busy. I'm on the job 6 a.m. to 8 p.m. I don't like to sit around, never did, never will, and this business ain't an easy one. You gotta love it. Always somethin' to do." He took the seat beside me and leaned forward, fully focused, with his elbows on his knees.

"So how can I help?" he asked. I felt like I was wasting his time with the background questions, so I tried to move through them as quickly as possible.

"What first brought you to Alaska?"

"Logging."

"Okay. Can you tell me why your work in logging is important to you?"

"That's all I know," he laughed, seemingly eager for more interesting questions as well. "I'm a third- or fourth-generation logger. My grandfather was a logger and a sawmiller sort of a guy, and my uncles and my dad, too. My son's in it now, too, and I hope my grandson can be a part of it." His family had come from Oregon to Alaska, chasing the promise of a burgeoning industry.

"We'll talk a little bit specifically about yellow-cedar trees, yellow-cedar forests," I said. "I'm just wondering whether you distinguish those trees and those forests differently from others."

"Oh, I'm looking hard for 'em. I mean there's only . . . there's hardly . . . Yellow-cedar is not evenly dispersed over southeast. There's patches of it here and patches of it there. Some places don't have any at all, and that is a significant thing. The cost of doing business in this end of the Tongass is staggering. We need yellow-cedar to keep us going. There are lots more stories to the whole thing, but fundamentally, there's an end comin' to yellow-cedar that's accessible. How do you get it when the last tree within reach off the road system is gone? It's not gonna be there."

Wes was talking more about access to trees than impacts from climate change.

"It sounds like a critical piece in your—"

"It's critical," he interrupted. "It's very critical."

"Can you tell me a little more about that?"

"Well, it's just, you know, the demand's high and the value's high. That helps to offset the low value of the other species. If you didn't have the yellow—a yellow-cedar component in the stuff that we do as a mill—we wouldn't make it. That's how significant it is."

With forty-five years of logging in the archipelago, Wes had lived through the golden age.

"Little people like us are the only ones left," he said. "In the old days, they were cuttin' five hundred million board feet. We're lucky if we saw five hundred thousand in a year. That's an immense difference."

His challenge now was getting more out of less. "We cannot survive just doin' rough-cut materials to ship out, to have it reprocessed. We have to do high-value-added products right here because we don't do enough volume. It's all about volumes and value. If you do a huge amount of volume then your margins are narrow, but you do enough volume to cover your overhead. If you're not doin' the volume then you've gotta go some other direction. Then you gotta go to value."

He was obviously concerned about the long-term viability of continuing to harvest yellow-cedar—more because of the waning industry and tightened regulations than because of the dieback—but I still needed to ask about the dying trees and whether he saw the standing dead as an opportunity or as a grave loss.

"Well, there's a theory that it's because of not enough snow cover over the roots in the wintertime," he said. "My observation is that there may be somethin' in that, but boy, I've seen years when there's been no—very little—snow, and other years where there's been just tremendous piles of snow." If climate change was taking its toll, it didn't make sense to Wes that some trees would still survive and that there would be places, in Hoonah or on the outer coast of Glacier Bay, where the living trees could still thrive.

Wes saw the standing dead as equally valuable, and perhaps even more so than the live ones. "If I can get to 'em," he said, "I'm happy to use 'em, but access to 'em is difficult and getting more difficult."

"Why would it have more value as standing dead?"

"Because there's an aura about using a standing dead tree for products."

"The idea that you're maybe not harvesting a live one?"

"Right. Right. Yeah. I'm sure you've run into that across the country."

Wes argued that to be useful, for there to be opportunity in the dead yellow-cedars, those trees needed to be harvested.

When I asked him if he thought of himself as connected to the yellow-cedar forests in any way, he burst out laughing.

"It's fine," I said, maintaining my composure and chuckling with him. "You can laugh at my questions."

"As connected?" he kept laughing. "Well, well, I'm connected because I need it to function. I need it to survive. That's the connection. We need some yellow-cedar. If I could figure out a way to grow it around here faster, or fast, I would."

On some level, his response made sense to me. There was certainly less cause for concern if he was skeptical that climate change caused the dieback, if he believed that the patches of dead were part of the natural cycle, and the species would ultimately regenerate elsewhere. If he could maintain the same relationship with a dead tree or a live one by getting the value he needed, his response was more about finding a way to innovate his business to adapt to the changing forests and logging industry than coping with loss.

"We're integrally tied to the forest," he told me. "It was put here to use and to utilize and to make useful for people. So, I'm tied directly to it. It provides jobs. It provides cover. I mean, it's totally utilitarian. I am not one that is spiritually tied to a tree 'cause a tree has no spirit. There's no spirit in a tree. God made 'em, created it for us to use, and that's my viewpoint on it. I am a conservationist, so to speak, but a practical conservationist. I want the trees to come back. Great. I want to have good, healthy trees, but to worship one? Uh-uh." Wes shook his head and grabbed his knees with his hands. "That was never meant to be."

SPRUCE, HEMLOCK, AND cedar lay in stacks surrounding his mill site not far from town. The road turned to dirt a few miles out. Instead of street names, numbers like "805" marked the logging roads. Wes said he knew the roads so well he could drive them with his eyes closed. The yellow-cedar logs outside the mill were painted with a yellow dot on one end, and he walked me through pile after pile, telling me what he would make out of each tree. Knots, burls, and cracks didn't faze him. Everything would find a home and a purpose.

"It's not just about clear wood anymore. I don't waste a thing," he said. He'd craft countertops and chests from the large logs, plaques and shelves from the smaller ones.

"This here is for tongue-and-groove paneling," he said, turning a smooth board over in his warehouse to reveal the distinct cuts for joining one to another.

Outside the building, he rested his hands on another stack of logs. "These are a few standing dead that came out of one sale."

In his truck, on the way back to town, he pointed out turns for the old logging sites and told me more about his family. Keeping his business going was an all-consuming endeavor, but I kind of envied his practical perspective on the changing forests and the changing industry. Doubting that climate change was the cause of the dieback almost made things easier, and so did his focus on use and economics. Being able to value the tree dead or alive purely for its functional values meant there were two courses of action: pressing managers to allow for more access to trees, and innovating his business to create a market for whatever he could harvest. There was nothing to process emotionally. No way for the death of the tree to trigger any fear about the future impacts of climate change, any feelings of helplessness, nor yearning for bigger, global solutions.

I thought back to what my father's friend had told me at his memorial—that our relationship to the deceased determines, in part, how we respond. Death can feel so cataclysmic in our lives, but only when we're close to what we're losing, when there's nothing that can replace it, and especially when the death is sudden. There's something about a gradual process, in contrast, that creates the space for acceptance or fighting, and ideally some combination of the two.

Wes's hours in conversation with me were hours lost from making his business work. Time shaved from the community of people that benefited from it. I wanted to give him a hug when we parted but went with the handshake instead. I didn't tell him I'd already interviewed people who would hike to a specific spot in the forest only to be in the presence of a live yellow-cedar—that those same people stayed clear of the dead, that there were some intangible values where substitutes could not do.

I began thinking there are relationships that are formed by functional needs but there are also ones created by the deeper emotional needs. For all the work we do in our lives, the more intimate and emotional realms may be the hardest at times. We can push the challenges aside, but that comes with consequences. I'd put my father's death on pause for all the interviewing I needed to do, but the reality kept surfacing.

How long can we put climate change on hold for the more immediately pressing? What will make denial no longer an option? When does a slow-burning ember become wildfire?

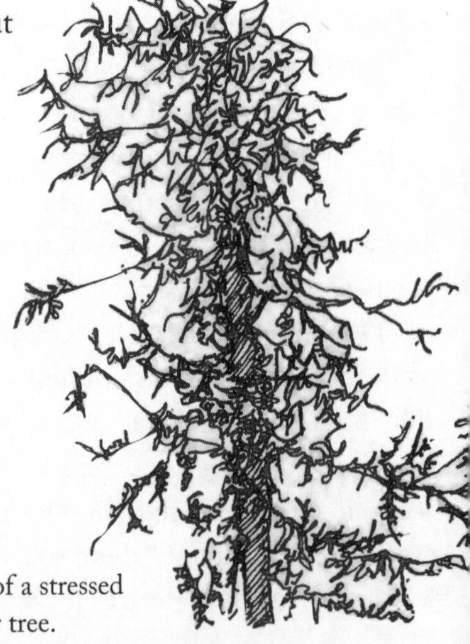

Maddog's doodle of a stressed yellow-cedar tree.

CHAPTER 8

Apart and a Part

I DIDN'T NEED to use a script for the questions anymore. Repetition had implanted them in my memory. Instead, my notebook made for a good table on my knees. I rested the recorder on top.

"My name is Kasyyahgei. I have another one. It's Kasake. Those are my Tlingit names," she said. "My English name is Ernestine Hanlon-Abel. I am Tlingit. I am Raven, Dog Salmon, Crow. We always follow our mom. My dad is Eagle Shark. I grew up in Hoonah. My dad is from Glacier Bay. My mom is from Angoon."

Ernestine wore a bright purple shirt with the faint outline of Hawaiian hibiscus flowers printed across the front. Her smooth, gray hair parted in the middle. I watched it bob, just so slightly, to the rhythmic delivery of her lineage.

Her house was cozy and colorful. It smelled a bit musty. A drawing of a raven and an eagle was framed on the wall—two powerful birds for two moieties of the Tlingit tribe.[1] With their heads pressed against one another, they looked in different directions.

"Balance," she said, noticing my stare at the image. "Everything is about balance."

She was the first to tell me that the forest gave her and her people their identity. Less than twenty-four hours after Wes had dropped me back at the bunkhouse, I was in her neon green home on the hillside in Hoonah, listening to her explain that the live yellow-cedar trees still standing were the ones to remember her ancestors.

A woman who worked in tribal relations for the Forest Service had referred me to Ernestine and another weaver, Teri Rofkar—two Tlingit women still practicing the traditional craft and passing it on to other community members in the northern part of the archipelago. On the phone, before we met, Ernestine offered her consent for an interview on three conditions. I wasn't allowed to ask about her weaving techniques or make any attempt to document them. (Researchers had come and done so before, she explained, then freely shared what belonged to her people, never quite getting it right.) Her niece and apprentice, Cathy, would join us, as a means of sharing stories across generations. Lastly, I needed to send her a transcript of our conversation to keep in the community what others typically only took away. I agreed to them all.

I told her that I wasn't an anthropologist; I wasn't searching for secrets into the art of Tlingit twining; instead, I wanted to understand how she related to the forest, to know what changes she had witnessed. Her request for Cathy's presence meant her time was a gift for more than me, and I appreciated that. But I also took it as a statement that her trust in me was tenuous, and that made me slightly uncomfortable.

Tlingit, Haida, and Tsimshian are the three main languages and cultural groups of the indigenous population in Southeast Alaska today. Scientists generally agree that people had settled the region by at least ten thousand years ago, after the end of the last major ice age, but there is little direct evidence of their ethnicity.[2] Anthropological archaeologist Dr. Madonna Moss wrote, "The evidence brought to bear on [the various] hypotheses has been limited to stone and bone artifacts, materials not often directly tied to ethnic identity."[3] Most Tlingit, however, say they've been in the archipelago from time immemorial. With a population of around ten thousand remaining in the tribe, the Tlingit are far more widespread in the region than the Haida and the Tsimshian, who migrated from their original homelands in British Columbia.[4]

If there is a community of people in the northern reaches of the archipelago today with the longest cultural history of using yellow-cedar, it is the Tlingit. Together with other Northwest Coast peoples of Alaska and British Columbia, the Tlingit, Haida, and Tsimshian practice traditional Chilkat weaving, which uses the golden inner bark of the yellow-cedar tree as a source of fiber. Stripped from a tree, the bark is exceptionally strong. Scientists call it the *phloem* tissue, and in a living tree it transports the sugars produced by photosynthesis in the leaves to other parts of the tree.[5] Native Alaskans have long woven it into blankets and clothing.[6]

"I really just want to understand your perspective of the forests," I had said to Ernestine on the phone. In her home, I said the same, again. If I was going to capture the range of ways in which people relate to the forests in the northern panhandle, talking with members of the Tlingit tribe was absolutely critical. I felt nervous. I felt fortunate to be there. I felt like I didn't want to mess it up.

The three of us packed into Ernestine's tight working space. For whatever reason, she had welcomed me in and given me a chance. I didn't want to frame a question incorrectly, or somehow misstep and end up like the others who had let her down.

I sat beside the wooden loom. It was a simple structure: one long beam on top with holes drilled through and a leg on each end for support. Woolly, bone-colored yarn and thinner threads of black dangled from the start of a new project. Books and loose papers cluttered the room's edges, making the most of the space for craft in the middle.

"We were the original tree-huggers," Ernestine said. "I don't care what they say. You can call me a tree-hugger anytime. The reason why we did that, when we went to the yellow-cedar tree, we put our arms around the tree." She raised her arms to embrace the air between us. They formed a circle as her fingers touched.

"If our arms went together like that, it was too small." She opened her arms, widening her hug, then looked upward as if a centuries-old cypress towered above.

"If our arms . . . didn't touch, the tree was just right. You could take the bark off. Yeah, we were the tree-huggers, the original. That's us."

Harvesting bark from the smaller trees would have killed them, but the bigger ones could endure losing a narrow strip. Ernestine didn't use the term, but I knew these trees as what researchers call "culturally modified trees," or CMTs. Whereas the inner rings of a tree can reveal how its growth relates to climate over centuries, the signatures left from bark harvests document a long history of people's use and occupancy of place.[7] The first CMT I ever saw was in Ketchikan, farther south in the archipelago. The tree, a red-cedar, had grown over its wound, softening the edges of what was once a sharp cut. I wondered who had been the one to create it. What was she wearing? What year was it? What was the climate like then?

"I think that in many ways, especially with the Chilkat weaving and the things that we do, we were truly the scientists of the world, of our world," Ernestine said. "We see what goes on, and we adapt for thousands of years."

When I asked her how she used yellow-cedar, she got up from her chair and squeezed behind Cathy to open a dresser drawer. She pulled out a bundle of light brown twine.

"This is cedar here?!" I questioned as I caressed the satiny strands in my hands.

"Yeah. That's cedar."

"How will you end up using this, weaving with this?"

"It's gonna get spun into my warp," she said. She reached inside the drawer again and presented me with a ball of soft, white wool—mountain goat wool, she told me.

"I lay the yellow-cedar in there, and I spin down . . . Spin, ply."

"It makes it stronger?"

"Yeah," she grinned. "Like I said, we were the scientists. It'll also keep moths away from it. Look at all that wool," she said, pointing to the thick, fluffy threads dangling from her loom. "Been spinning that millennia ago." The cedar kept the wool firm.

I relaxed a bit, feeling some common ground; there was shared value in making observations of nature and recognizing the ways in which people rely upon it.

She reached into another drawer and then unfolded a finished weave before me.

"It's an apron," she said, wrapping the heavy cloth around her waist and tying the two ends behind her. Black eyes and yellow faces stared back at me from the weaving. A blue shape in each corner resembled whale tales about to slap the sea. She swung her hips and began singing softly in words I didn't know. A chant for just seconds, ending with "Hoo haa! Hoo haa! Hoo haa!" as she flaunted the masterpiece on her body.

"From here to there," she said, running her fingers over a section, "I worked with a hundred and nineteen pieces of yarn."

"And there's yellow-cedar in that as well?" I asked.

"Yes."

"That is amazing."

"See the circles?" She touched one at the top and whispered slowly. "See the circles, ciiiircles." She raised her voice again and declared, "It's the only weaving in the world with perfect circles. No other weaving can even come close."

I felt uncomfortable again, as if we were on the edge of violating term #1 (no inquiries about weaving techniques) without me even asking, so I moved back to my usual line of questioning. But the look on her face as her forefinger traced the circumference of each circle reminded me of Wes's whenever he'd laid his hand on a slab of yellow-cedar at the mill. There was a sense of pride, but also one of intimacy.

"Could you tell me the ways in which you value a yellow-cedar tree? How do you explain that?"

"Ooh, help me," she said, sitting back down in her creaking chair and turning to Cathy. "She's my good English person."

"It would be nice to have time to think about that answer," Cathy replied, "because you're asking, what's the value of your culture? It's the same thing. They're interwoven. How valuable is Alaska Native culture? That takes time to answer."

"If we don't have our trees, we don't—we can't be who we are," Ernestine said.

I wriggled in my seat for a moment. *Does she know? Does she know about the dieback? The graveyards? The massive scale of dead trees on the landscape and the species' uncertain future?*

I dreaded the inevitable follow-up question. I didn't want to hear their pain, their loss, if they knew they were losing the tree, and I hated the idea of holding back what I knew if she didn't. Holding back felt dishonest, but sharing my own knowledge would have shattered any chance of understanding hers (or that of anyone else I interviewed). I felt like a medical doctor emerging from the operating room with bad news, except I wasn't allowed to deliver it. I had to trust the research process; this was how it had to go.

When I asked about disturbances and impacts to yellow-cedar trees from human activities—a question that commonly incited talk of climate change—Cathy and Ernestine both pointed to logging. Only with more questioning did the graveyards emerge. They'd seen patches of dead yellow trees, but only a few. From all the interviews I'd done so far, it was already evident that learning about the dieback came partly from direct exposure to the dead stands. So, if someone happened to live near healthy trees, and hadn't traveled to an area where the dieback was more extensive, the large scale of the affected forests could go unnoticed. When I interviewed them, neither Ernestine nor Cathy identified climate change as the culprit. Ernestine thought maybe the Forest Service was spraying weed-killers to control the forest understory.

"I was curious about Roundup because it was just one selective area . . . off the road," she said about one of the smaller patches she'd seen.

I asked the same question another way. "So you haven't heard too much about those areas of standing dead? That there's something affecting the yellow-cedar?"

"Logging," Ernestine replied. She kept pointing to the history of logging as the main source of concern for the loss of yellow-cedar trees and apologized, preemptively, for going off topic.

She launched into a story her father had told her about a man who came from another village to see him. He was running for the state legislature. Her father told him to sit down; he wanted to tell him something. "He says, 'See how the mountains are? A lot of avalanches, huh? Yeah, there's lots.' He says, 'I'm gonna tell you what I need from you. You're gonna have to learn how to hold hands the way those trees do. The roots, they send out all these roots, and they learn how to hold hands. As they hold hands, every year, you see a little bit more come in, a little bit more come in, a little bit more come in. They're joining hands. Pretty soon, the avalanches aren't gonna be able to go down there anymore and wipe it out. That's your job, to hold hands.'"

She smiled at me, waiting for me to make my own connections, and Cathy nodded.

"We were able to use analogies," she said. "It was also our tool for teaching and helping us build the community." Ernestine never said so directly, but she was asking me to consider community and collaboration as our strengths in the face of disaster. She was focusing on clear-cuts. I was thinking about climate change. In either case, individuals coming together may be a means of resolving whatever imbalance we've created.

"The forest gives, gives, gives, and it's our job to learn how to give back to it."

I thought back to Wes. *What did the dieback mean for Wes?* Wes Tyler could adjust. He could substitute a dead yellow-cedar for a live one, or find other uses for a spruce. *What did it mean for Ernestine?* The more I talked with her and Cathy, the more I could see there wasn't anything that would replace the value of yellow-cedar for them and their people. That inevitable loss—the result of different values and ways of relating to nature—felt unfair to me.

"We communicate," Ernestine said. "The trees and I, we communicate. They talk to her, too," she added, pointing to Cathy. Then she looked at me again. "It's just you have to learn how to listen."

"Working with cedar, mentally, you're connected to the Earth the whole time you're touching the weaving," Cathy said. "You're connected to your ancestors at the same time."

"It becomes a spiritual connection," Ernestine added. She described the times she felt most spiritually connected as when she was weaving. "All the designs," Ernestine said, "show something from the land."

I placed the recorder on the floor to make a few notes about how Ernestine valued the live, sacred cypress:

Who we are
Our culture
Healing
Feel happy, mental
Connection

For the standing dead, I wrote only one word:

Firewood

OUR CONVERSATION TOOK a twist when I mentioned the parks and wilderness areas set aside to protect trees and other natural resources. I knew these kinds of designations weren't stopping the impacts of climate change, but they'd halted logging and other direct forms of human activity in some places.

"We're not allowed to hunt the mountain goat in Glacier Bay, but that was our traditional hunting ground," Cathy said. "When you say, 'set aside,' it's like, 'hey!'," she gasped. "If you put the question to me, like, 'Will you agree to "set aside" lands if that stopped the logging?' Ahhh. I wouldn't. I wouldn't."

Ernestine and I reacted at the same moment, inadvertently talking over one another. I thought I heard her say, "Wilderness is a curse word."

"Say that again?" I couldn't believe I'd heard her right. Years ago, I'd fought for the conservation of lands threatened by mining in Native communities. I'd supported organizations aiming to protect the watersheds of rivers that I once guided. Today, under climate change, I think we need to consider how we manage our nation's protected areas more than ever before—and we need to decide what values (or services for people) we want to sustain.

"Wilderness is a curse word to us," she said again.

"I would rather educate the people and see—have them learn the value of what they're using, as opposed to set aside and make it all stop," Cathy added. "Because once you've set that in motion, it becomes political football. Who wants to rape the land the most? We're a minority here. We're a small voice trying to say there's true value in this land. This is one part of it that you're checking, the cedar."

I said nothing, but a quotation I had pinned to my office wall at Stanford popped into my head: "We simply need that wild country available to us, even if we never do more than drive to its edge and look in. For it can be a means of reassuring ourselves of our sanity as creatures, a part of the geography of hope."[8] It was from a letter that the writer and environmentalist Wallace Stegner had sent to a man named David Pesonen at UC Berkeley in 1960. At the time, a research center at the Forestry School was under contract to assess the status of the nation's wilderness resources.[9] Four years later, Congress passed the Wilderness Act—creating the highest designation for preservation that still exists today. The act poetically called wilderness "an area where earth and its community of life are untrammeled by man, where man himself is a visitor and does not remain," and it raised the bar for protections in our country.[10] According to Stegner, we needed to "put into effect, for its preservation, some other principle than the principles of exploitation or 'usefulness' or even recreation."[11]

Stegner, and many other activists, saw wilderness in the United States as the last islands of hope—the pristine areas of "nature" to be saved and set aside. Wild places were refuges from the grind and congestion of urban life. Woods offered a contrast to humanity. I'd read other scholars who had critiqued the notion of nature and wilderness as separate. In his seminal environmental text *Wilderness and the American Mind*, Roderick Frazier Nash concluded that wilderness doesn't exist. It never has. Instead of a physical reality, it's a state of mind.[12] Environmental historian William Cronon called wilderness "the product of civilization" and argued that "we mistake ourselves when we suppose that wilderness can be the solution to our culture's problematic relationships with the nonhuman world."[13] But to sit there in Ernestine's home by her loom and hear her call wilderness a curse word, to claim the designation itself is to blame for the imbalance we've created on our planet, that struck me.

Our separation from nature stems from our early efforts to protect it? And that separation is the cause of our problems today? There was an irony and unexpected twist—the once well-intentioned act of protecting wild places had broken the relationships needed to sustain the larger whole over a much longer time frame. It was the exact opposite of what Stegner and the National Park Service would have wanted. What I had once fought for, she was fighting against, but I didn't feel defensive. I felt like I had something to learn.

"If I can really understand that whole perspective—of you being a part, us being a part," I questioned, "then you're saying *that* is what should stop logging?"

"It should. It should," Cathy said. Formally designating lands as "Wilderness" had severed the relationships she and her people had cultivated with the natural world. Ernestine said their relationship to the land and trees had always been one of balance and respect. Just as a curse word divides two people in conversation, setting aside nature tore it apart from humanity. For someone like Ernestine, drawing lines on a map and preserving places for people to connect with "nature" made no sense. This approach was only logical to people who had lost that connection and already severed relationships.

I could see Ernestine's argument. Designation made nature something we reach by driving, or, in Alaska, by flying and boating to an edge. Drawing lines on the land to identify some places as wild or pristine meant we could do anything we wanted in other places, because those islands on the landscape—Stegner's pockets of hope—preserved the ideal. Meanwhile, in the half century since the passage of the Wilderness Act, we had burned our fossil fuels with abandon, realizing, too late, that the environmental effects of industrialization would know no boundaries. Park or urban center, public or private, there is no land designation that protects a place and its people from climate change.

Cathy said, "The education has to start somewhere. It has to start at some point. The fact that you asked the question is a beautiful thing. You're one out of a million." Ernestine nodded, offering me approval. I didn't feel like one out of a million. I felt like my disconnected culture was to blame, that somewhere along the way my own ancestors had gotten off-track in the race to industrialization. Maybe she could sense that I wanted to help fix that, to get the whole back on track and in balance again.

I had to tell them about the tree death and its cause before I left, after the interview was over. I didn't know if they'd believe me, but I didn't think the cause would even matter to them. The solution was the same, whether the cause was death by pesticides, death by logging, or presumably even death by climate change. Moving forward toward something more sustainable, toward balance, required seeing nature as a part of humanity, seeing one in the service of another as a whole.

I explained the patches, the shallow roots, the long history of studies Paul Hennon and others had conducted, my own work on the outer coast. Cathy placed a hand on her heart. Ernestine held her composure constant. It was the most I had spoken in our entire visit.

I was getting to the end of my explanation. "A lot of my research has been trying to figure out in those areas where there's a lot of standing dead—"

"There's been changes," Ernestine said before I could finish.

"Yeah. Do we see regeneration or—" I was going to tell her about the decrease in saplings in areas affected by the dieback.

"The climate has definitely changed," Ernestine said. "One of the things her grandma did, and my aunt, was make these four-season flowers." It was a Tlingit pattern with four distinct designs for four distinct seasons. "We had real defined seasons. Our seasons are not defined like that anymore."

"We may get snowfall, we may not," Cathy said.

"It may rain all winter, and like all summer, last summer, it just rained," Ernestine shrugged.

"It's called the four-season flower because it's a white flower to represent winter, a green flower for spring, a pink flower for summer, and a brown flower for fall, or a brown leaf. It's just all beaded into the pattern that depicts our four seasons. But they're not as defined as they used to be."

What was there to say? I had nothing. I nodded.

THERE IS A story about the origin of yellow-cedar told by a woman named Alice Paul of the Hesquiaht people of Vancouver Island in British Columbia. Paul Hennon had sent it to me around the time I had transitioned from entering plant data to formulating interview questions.

Raven, known as the great creator and trickster, came upon three young women drying salmon on the beach. He wanted their salmon, and so he kept asking them if they were afraid to be there by themselves—if they were afraid of bears or wolves or other animals. When he learned they were scared of owls, he hid in the bushes nearby and began to imitate owl sounds. The women fled, running partway up the mountain until they were too tired to go farther. They stood on the mountainside to rest together and turned into yellow-cedar trees. As the legend goes, that is why yellow-cedar trees are found on high slopes and why they're so beautiful.[14] In other Tlingit stories, the raven is thought to be an alarm for all creatures. When the raven crows and a nocturnal creature is still away from its home, the creature dies instantly.[15]

I was a few steps from Ernestine's home when a raven called out in a series of loud caws. The bird swooped down from a spruce tree in front of me, and I could feel the wind from its flapping black wings on my face. I watched it disappear into the forest.

Maybe it wasn't a canary I was chasing anymore but a raven. Was the raven issuing a call for alarm? Was the birdsong a warning? Or was it a call for something more than hope—a knowing that what disappears in one place, like the three women on the beach, may turn up somewhere else—that what we lose is not always gone forever?

Before I disassembled my mountain bike and handed it, in pieces, to a pilot flying me out of Hoonah, I interviewed a master carver. His name was Gordon Greenwald. His father was Tlingit, and his mother, from Colorado, was not. He and Owen James, a full-blooded Tlingit, had just begun carving a totem from yellow-cedar that would ultimately go to Glacier Bay, the place where the ancestors of the Hoonah people had lived before the advancing ice of the Little Ice Age had forced them to resettle.

"If our climate is changing," Gordon said, "then our environment will change, and that doesn't mean as humans that we can't change with it."

I didn't think of him as just hopeful, because there was a sense of conviction in his voice. As a scientist, I couldn't measure that or categorize it into what I was studying—the K-A-B; knowledge about the dieback, attitudes about the changes, and behaviors in response to the standing dead simply didn't apply. It was a way—his way—of waking up in the morning and living in a world of loss and change. Where there is tragedy. Where there is beauty. Where our actions in coming together and working together to reshape the hillside can still stop the avalanche. Where finding balance again is within the realm of possibility. With my father fading into memory as I continued traveling from interview to interview, home to home, town to town, that faith began to be enough for me, too.

TERI SMILED AND gazed across her porch to the trees beyond. I checked the dials on my recorder and repositioned the microphone to avoid the wind. It was late May. I had conducted three dozen interviews so far that spring. She straightened her t-shirt around her waist, brushed a loose strand of long, gray hair behind her ear, then returned her eyes directly to mine.

"My name is Teri Rofkar, and in Tlingit it's Chais' Koowu Tla'a. I'm T'akdeintaan, a Raven from the Snail house, and I am the daughter of an Englishman and the granddaughter of Kaagwaantaan, Wolf. Our homeland was Lituya Bay, Glacier Bay area. That's where our clan group comes from. We're very closely related to the L'uknax Adi, the Cohos. Those are our folks."

She slid one hand over the other in her lap, and the long sleeve on one arm slipped up. I noticed a tattoo: one thick, solid, black line running across her wrist. To me it appeared like a symbol of boldness and strength, fitting my first impression of her.

"I am a weaver. I weave full-time," she said.

Curious to see her work and thinking a slow start could increase her trust in me, I accepted Teri's offer for a tour of her studio before sitting down to record our interview. We marched up a slender staircase and slipped around the top corner through a small, creaking door. Inside, shelves of books lined the perimeter. I scanned across their titles, trying to discern a pattern in the rows—Native culture, Tlingit and Haida, Alaskan history, botany, art, Eastern philosophy. On the wall above one shelf, she had printed a quotation on a piece of square paper: "*If you are depressed, you are in history. If you are anxious, you are in the future. If you are at peace, you are present.*" A large loom filled the far end of the room. White yarn, like the mountain goat wool I'd held in Ernestine's home, dangled from the start of a new project.

"I collect it in the woods, and sometimes I trade for it. I weave in spruce roots."

Handwoven baskets stretched the length of a long table. A three-ring binder rested open, turned to a page with photographs of Teri in a striking wool robe. Her arms were spread wide open like the wings of a raven, exposing the full design in its grandeur: black, angular lines and turquoise trim, tassels hanging from the edges.

"The cedar are impacted by the climate change," she had told me on the phone, "so I just can't see still harvesting even bark the way we used to do. The blessing of the cedar is that it single-handedly retained the weaving for so long. But now we need to give it a break."

Teri passed a small basket to me. "This one is from cedar and spruce." I ran my fingers across its woven surface, touching the smooth and sinewy bark—still fragrant. The sweet smell of cedar carried me from her Sitka home to the outer coast.

"This one is from spruce."

I waited for whatever she shared next.

"Let's go to the porch," she suggested.

Outside, with the recorder running, everything seemed to be going smoothly until I used the word "resources," as in "natural resources," in a question. That lit a fire. She leaned toward me.

"One of my goals," she said even closer, "is to eliminate the term 'natural resource.' I think that it's just—it's an atrocity—you know, it's the resourcing of everything. There's no relationship. If you just replace the words 'natural resource' with 'relationship,' you're good to go. When we resource, we don't make the ties of what was lost in order to gain something."

I made a quick mental checklist: *The plants I eat; the water I drink; the wooden roof over my head; the mined ore for the metals that go into my car or my computer and the microphone in my hand. Natural resource management—there are advanced degrees and government jobs to master just that. We are resourcing absolutely everything. The fossil fuels I use.*

I felt an upwelling of guilt for my own "resourcing." *How much slack am I willing to give myself? How much slack should I give myself?*

But, emotions aside, scientifically, if I had a relationship with everything I resource, what would that change?

I decided to stray from the topic, avoiding any use of the resource trigger word until I could find a way back. I asked about her observations of climate and changes in the forests where she harvested.

"We're looking at things all the time with research in mind. Others might say, 'Oh, I'm doing science now. Wup, I'm just recreating now. I'm doing science now.' But when I go into the woods, I'm doing both." Teri told me everyone has a responsibility to be alert and aware and to respond, in our actions, to the changes we see.

"My spruce roots made a change eighteen years ago," she said. I was intrigued. "It's kind of a physiological change in how the root system I work with works, and so that's something I'm monitoring. Yeah. I don't know how to articulate it."

"The roots dive deeper?" I probed.

"No, it has to do with some sap and the way they groove. They used to be all lumpy, and that was when you stopped harvesting. They don't do that anymore. It's like the window has changed. They harvest roots year-round down in California. We're getting closer to being able to do that. The root system changed—the way they grow."

I asked her about uses of the forest. She said her spiritual enrichment is just as valuable as any food she walks away with. We wound our way back to relationship, not resource.

"You can create balance," she said. "You can manage balance. Here we cut down an old tree, and now there's a new tree coming up. Look, that's in balance. So, when I say relationship, it's relationship with the whole, not the individual. It's relationship with place. Place-based learning. That's hard, because I think so many people are disconnected with the place where they live. We've got this superficial layer—whether it's technology or transportation or whatever it is—but it's created this little isolating layer from place. I think people get it when they garden, but, again, you're in control of that situation. When I go out in the woods, I'm not in control. I'm not the top of the food chain. I'm a visitor. It's a relationship. A different relationship." She threw her hands up in the air.

"What do we know about relationships?! They are messy!" she exclaimed. "And difficult! Oh, there's more to that word than I'm giving you!"

Almost picking up the conversation where Ernestine and I had left off in Hoonah, I chimed in: "There may be times when that relationship is thrown out of balance."

"Yes!" she nodded. "And times when we figure out how to balance again. But I'm willing to engage again and say, 'Wow! That really didn't work out. What was I thinking when I took all that bark in one area of the hillside?'"

I thought of my dad, and the times he'd been there, the times he hadn't. I thought of restoration in relationships—the actions people take to repair the environment, and the actions people take to support one another, long-term.

Teri seldom responded to a question with what I'd call an answer. It was my job to decipher meaning from stories and more questions. Replacing "resource" with "relationship" would make nature and humanity a part of one another again. A relationship is so much more than a service provided or a resource to use. It is a mutual commitment to care.

THERE'S A POINT in the interviewing process that I was awaiting—social scientists call it "saturation point." Eric Lambin had mentioned it to me before I left for Alaska. Otherwise, I knew the term only from chemistry: it's the point at which the greatest possible amount of a substance has been absorbed into another. No more solute can be dissolved into the solution.

"You're studying knowledge of the dieback and its causes," Eric had said. "You're assessing people's attitudes about it. You're exploring their uses and values of the forests to determine the ways in which they may or may not be attached to the tree. And then you're looking for behaviors—any kind of response to the changes in the environment. Right?"

"Right," I confirmed.

"Well, at some point, twenty, thirty, or maybe forty interviews in, you'll be in the middle of an interview and realize that you can predict the rest of it. Not the nuanced personal information and experiences, but the broader outcomes you're studying. If there's a relationship between knowledge, attitudes, and behaviors, if there's a role that attachment plays in determining how someone will respond to the dieback, you'll see it. Eventually, after enough interviews, assuming you've captured the range of all those variables in a big enough sample size, you'll see it. Relax. Take a day off. Then do half a dozen or a dozen more interviews to keep confirming. If you're right each time, and nothing new pops up, call it good. That's the saturation point. You've got what you need for analysis."

Halfway into my interview with Teri, I paused for a few moments and went through a checklist of what I'd learned so far in our conversation.

In terms of knowledge, she knew about the extent of the dying cedars throughout the archipelago, and she knew what was causing their death. When it came to attachment, the functional and emotional were deeply intertwined for the many uses and values she derived from live yellow-cedars. Part of what she gained came from the bark, as with Ernestine—and from the wood itself, as with Wes. Then there were the intangibles, the values coming not from use but from interaction and existence. But could I predict her response—the ways in which she was coping, or any changes in behavior?

"Tree people," she said. "We have conversations daily. I love 'em. It's love. It's just love. I do put myself in the presence of cedar on a regular basis. You almost have to—they're all such different personalities, you know, so you have to go spend time with them on their own."

"Do you value an old-growth tree differently than a young-growth, and then in what ways?"

"It's not an economic value," she said. "The stories and just the history they've experienced. How do I measure that? There's not a lot of economic value in either one of those things, but as far as who we are and how we're connected, huge."

I didn't comment; it wasn't my place to do so. But I felt like my interviews were capturing what I couldn't measure as an ecologist, and I trusted that.

She has the most to lose, I thought. *The deeper the relationship, the more there is at stake.*

Her knowledge and her attachment to those trees should have meant she held a great deal of concern, and she did. She was worried not only for the dying trees and the loss of those irreplaceable values, but also for what else would come at a larger, more global scale.

"There are the amphibians," she said. "There are the cedars. There are those in our ecosystems that are the ones to warn us, and I think that that's what they're doing right now, and they have been for quite some time."

I could predict that concern would be intensified by knowledge of the dieback and attachment. Teri fit the bill. And if someone knew the cause was related to climate change, that could trigger another suite of attitudes, such as the fear of future impending impacts, or a sense of hopelessness.

But I still couldn't predict B, behavior, the response. I was close to saturation point but not there yet.

She has three challenges in front of her: changing her use of the live trees, coping with the loss of what she cannot replace, and confronting the enormity of the cause.

She called them the tree people.

My father's sudden death had hit hard with shock. Teri's loss was gradual. Like a cancer spreading, a slow death meant time to savor, time to accept, time to relate to one another differently, time to anticipate, time to prepare for the world without. But still, she was left with a gaping hole. She was losing her father, her father's father, and so on, back through time. It was a scale of loss unfathomable to me, now embodied in a tree—worse yet, a forest.

What healing can happen in the absence of the trees?

I'd had a yoga teacher back in Santa Cruz who used to say, "Fear is the absence of breath." There was nothing scientific to that, as far as I knew, but she had certainly done her work around the emotion. She'd say that we rationalize and construct excuses when something, or someone, asks us to step outside of our comfort zone. Action, she claimed, is what cures fear. She said fear should be your compass. It can guide us toward survival.

Researchers have conducted extensive studies and written books on the topic of fear and hopelessness in the fields of climate change education and communication. As awareness of climate change increases, why aren't we taking the actions needed to curtail it? Sociologist Kari Norgaard spent time in a rural community in western Norway interviewing community members during an unusually warm winter. In her book *Living in Denial: Climate Change, Emotions, and Everyday Life*, published in 2011, she wrote, "Although some 68% of the population [in the United States] list global warming as a serious environmental problem in recent polls, few people spent time writing or thinking about it, much less taking action." She explained that "not wanting to know about climate change appeared to be related to the host of powerful emotions the topic engendered. The people I interviewed described fears about the severity of climate change, of not knowing what to do, that their way of life was in question, and the government would not adequately handle the problem."[16] Living in denial, turning your back on fear, is, in many ways, just easier.

So, behavioral response (the B) remained the most uncertain variable to me. Why do some people freeze up, look the other way, and stick to the status quo? Why do others have the courage to try something new? I was starting to see that if we respond with forward action and a positive outlook, fear could be a part of learning to thrive.

I asked Teri more questions.

"Some of the areas that we mentioned affected by the decline, where there's lots of standing dead—I'm wondering if you've ever walked in those or—"

"I have not, and that's not by accident," she laughed. "I have not sought those places out. They make me uncomfortable."

"When you think about those changes or what you've seen—going to Peril Strait or wherever—what do you feel about that? Is there any reaction?"

"Sure. Just loss." She shook her head. "I don't look at the economic opportunity, but if I am indeed looking at a more sustainable or subsistent kind of a relationship, it would make sense that these dead trees are presenting themselves, and we should probably harvest the wood, if we can do it respectfully."

"Has your knowledge of the standing dead trees changed your use of yellow-cedar or those forests in any way?"

"Yes. I am an advocate for people to switch to spruce roots. There's a sustainable way to harvest them without jeopardizing the quality and the life of the tree."

She could substitute spruce roots for cedar bark, and that was the choice she'd made. Refraining from using the cedar, substituting other species, and avoiding the forests affected by the dieback—those were some behavioral changes she was making in her own life. Although none of her actions alone would stop climate change, or save the yellow-cedars, they constituted something different from the status quo. As for the loss of the tree people and the ancestors those trees held close, that part of Teri's response was more psychological. It was about grief. It was about reconnecting as a way of repairing.

At the end of every interview, there was a series of questions I asked that were never used in my analyses. (Like the typical researcher optimizing "field time," I collected more than I could use later.) I'd added the questions after the pilot interviews, soon after Greg Streveler left me feeling like hope was useless, that we are better off living a life outside of hope. But I didn't think that acceptance or giving up could be the only alternatives. I believed there was something more than just hoping for change—that we could (and still can) be a part of the change—and I wanted to find evidence of that.

The first of these post-interview questions was, "What is the environmental issue that concerns you most and why?" The second was, "When you think about this problem that concerns you, what's your outlook—there's not much we can do; there's some things we can do; I don't know if there's anything we can do; there's a lot of things we can do?"

Teri said the dearth of place-based knowledge was the environmental problem that concerned her most; most people lacked a relationship with the place they came from, with who they were. Others I'd interviewed had listed ocean acidification, pollution, global warming, and similar issues. A few had listed our lack of connection.

"There are some things we can do, and it starts with us," Teri said. "It's not the big picture, 'Oh my god, those guys over there, they are the problem—'"

I showed her a figure of a series of circles in pairs. In each set, one circle was labeled "Nature" and the other "Self." Across a scale called the Inclusion and Nature Scale (INS), they overlapped to varying degrees, from barely touching to totally overlapping. Results of studies using this measure show a correlation between connectedness and biospheric concern—a concern for all living things.[17] Researchers have also found a correlation between connectedness and self-reported environmental behaviors such as recycling, conserving water and energy, and using public transportation—the seemingly small individual actions that some doubt can make a difference.[18] I asked her to tell me which set described her relationship with nature, but I already knew her answer.

Where pessimism resided in numbers and future projections, perhaps optimism could reside in the relationships we create.

"It's such an odd thing to think of being separated," Cathy had said, when I had presented her and Ernestine with the same image. "People who arrogantly think they can damage our environment and not get affected by it, it's just a time bomb ticking."

"My dad always said we are a part of the ecosystem, and I'll stand on his word," Ernestine had said. "We are a part of the ecosystem."

Teri reached out and touched the image on the far right—the one with the two circles completely merged.

"That would be me," she confirmed.

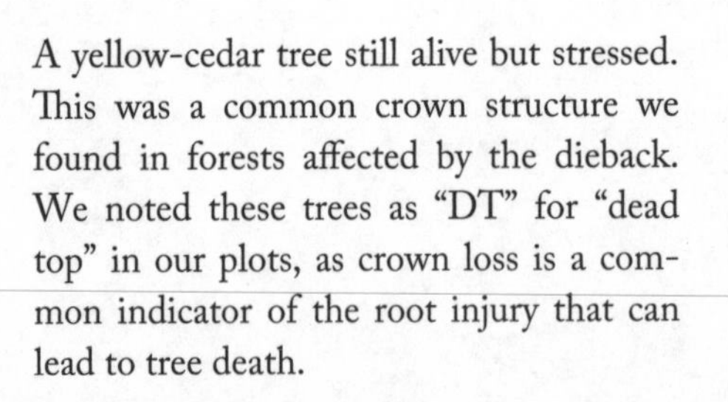

A yellow-cedar tree still alive but stressed. This was a common crown structure we found in forests affected by the dieback. We noted these trees as "DT" for "dead top" in our plots, as crown loss is a common indicator of the root injury that can lead to tree death.

CHAPTER 9

Saturation Point

EVERYONE IN MY cohort of graduate students took the Myers-Briggs Type Indicator test during our first year at Stanford. Businesses and consulting firms use the MBTI to build stronger collaborations between people. Counselors use it to advise on appropriate career paths.[1] It's based upon Carl Jung's theory of psychological types and four dichotomous preferences in orientation.[2] In the first three, we are each either extroverted or introverted; we perceive by sensing or intuition; and we make decisions by thinking or feeling. In the fourth, we approach life in an organized, structured manner, which the test calls judging, or in a more open, flexible manner, termed perceiving.[3] As I understand it, it's unlikely for your type to shift dramatically over the course of your life. Some of us may waffle between the preferences under various circumstances. But as a general rule, we are who we are. Once you know your type, there are some fairly predictable patterns in how you move through the world and relate to others.

The tests we took indicated I am an INFJ (introverted, intuitive, feeling, judging), and reading the descriptions of my type online gave me a surprising sense of relief. The portrait of an INFJ so clearly aligned with my own sense of self. Deep connections in my relationships are important to me. Values and principles have always guided my decision-making far more than money. At various points in my life, I've lived in a cabin without running water, a toolshed, a tent, and my car; all were cheap living choices that allowed me to pursue what I thought was meaningful work despite little pay. My ability to empathize means I can easily understand other people's emotions, but I also carry this sense of responsibility that what I do in life should alleviate inequities and injustices, not contribute to them. For better or worse, I relentlessly judge myself on that metric and constantly see the trade-offs.

My interviews with Teri and Ernestine made the injustices of climate change even more apparent to me. I could grasp the loss of my father; I was experiencing that. But the grief they experienced for the loss of a culture, for the links to their ancestors, for the "tree people" as equal members of a community and family—that grief carried enormous weight. Global changes are lived locally. I couldn't measure the sorrow I felt in Teri's presence. But sitting on her porch with that emotion made what I knew intellectually about climate change and the disparities of its impacts a more personal reality. Interviews with people in the midst of other environmental problems—farmers in California during the drought that year, for example—would have exposed other disparities. To me, the drought meant shorter showers and less watering. To them, it meant their livelihoods. Yet still, some farmers were switching to drought-resistant crops and integrating what we know about climate change into their practices. Some people find a way to change along with the changing environment.

Teri, Ernestine, Wes, and many others—they were losing something they relied upon in various ways. But I was beginning to see that loss could also lead them to develop new relationships with the surrounding ecosystem and perhaps beyond. The people most attached to the natural world—whether it was functional or emotional—could also be the first to adapt.

LATER, ONCE I'D conducted all my interviews, I would ultimately test whether different combinations of knowledge, attitudes, and attachment to nature could predict a behavioral response to the dieback. I expected those dimensions to function in ways that were similar to the

Myers-Briggs dichotomies, except that in this case, people could move fluidly between knowledge "types" by learning more about the dieback and climate change. I would separate my interviewees into groups based on what each person knew about the dying trees. I planned to categorize their attitudes by identifying concern (because concern can be a motivating factor for behavioral change) and whatever else emerged. Then, theoretically, I'd see a set of associated behaviors—like planting trees, refraining from using live trees in response to the dieback, or even biking to work and reducing home energy use. The latter would indicate that understanding climate change impacts in place and experiencing it as a direct threat could motivate people to reduce their personal contributions to the problem. In the scientific dream world, I would end up with some sort of prescription for motivating action that could have both local and global benefits. Even more positive change could come from other people fostering the "right" combination of knowledge, attitudes, or attachment (whatever "right" ended up being) to incite action at a broader scale. Voilà! Problem-solving, relevant research! We have citizens adapting their behaviors to the changes in their local environments and engaging in a larger movement to halt the climate trajectory!

The prescription turned out to be messier than I anticipated. Behavior was one dimension of a response. But there was an equally important component I never expected: coping psychologically.

AFTER I LEFT Teri, I began playing the saturation-point game for all the interviews that followed. Halfway through the questions, once I understood the interviewee's use of the trees, the values he or she attached to them, and how knowledgeable he or she was about the dieback, I would try to predict the rest.

My thought process went something like this: *So* ______ [insert name here] *knows the trees are dying and holds some level of attachment to them. Is he (or she) concerned? Predictably yes. Is that concern about the local losses? Predictably yes. Oh, but wait, this person knows climate change is the cause. Is he (or she) concerned for what this tree represents? Predictably yes, there's concern at the much larger, global scale. Is that concern for humanity? For the relationships that sustain us?*

In Sitka, I stayed in a dorm room at another Forest Service bunkhouse. Volunteers came in and out each day with reports from stream surveys and news of bear sightings. I stayed up late going through transcripts from interviews. I had sent each audio recording off for transcription as I moved from town to town, so the documents came trickling in whenever I got Internet access. I would spend a year analyzing them all systematically back at Stanford, but the ones I gravitated toward immediately were from the people most knowledgeable about the dieback.

How did they know it was caused by climate change?

What did (or didn't) they do with that knowledge?

"With the dead standing in the forests and what's going on now, I think we're dealing with a bigger picture than you or I are ever going to be able to—we gotta stop driving so many cars," a Tlingit carver from Haines told me. "We gotta plant more trees. We gotta slow down. This electronic age has got a hold, and where's this going? What's it gonna do? Does anyone ask, or are we just diving into this more and more and more? It seems like what will happen is Mother Nature is gonna throw a fit."

Hank Lentfer, a conservationist and writer in Gustavus, said, "Part of loving this place is coming to terms with the grief associated by what those cedars represent." He argued that professors of ecology and environmental studies "need to integrate the psychic component on 'How the fuck do you deal with this?'" so we "can go on and formulate a positive response to it."

There was the boat captain, working for a local tour company, who said the decline teaches us a critical lesson. There was the naturalist guide who said if it's an indicator of us not taking as much care as we should or could, or not considering how to keep our ecosystems whole, then that was what was important. There was the conservation advocate in Juneau who said it was a red flag—a flag that resonates to stress on the broader ecosystem. I listened to her audio recording over and over again: her pace accelerated, her tone became emphatic when she asked,

"Does this motivate people to realize what's at stake? We're visually seeing our fears manifest themselves."

The boat captain called the wild country "a crucible of what the Earth is doing."

Observational ecology and social science share the common goal of discovering patterns. And patterns are only revealed through repetition. I asked the same questions again and again. After more than forty interviews, I could place my participants into one of four groups I'd identified on the fly regarding knowledge: (1) the people who didn't know about the dieback (there were only two in the end; both of them lived near healthy trees and traveled infrequently); (2) the people who were aware of the dieback but not its cause; (3) the people who were aware of the dieback and its link to climate change, but limited in their knowledge about the climate mechanism; and (4) the most knowledgeable—the people who were aware of the dieback and the intricacies of the killing culprit.

The most knowledgeable told me about the loss of snow, the roots freezing in cold snaps, and the higher risk at lower elevations.

"When we do get our colder temperatures," a logger from Tenakee Springs said, "those trees are more subject to freeze-out and dying from their roots not being insulated by snowpack."

The people in the most knowledgeable group were the forest managers, who, like me, visited the graveyards through their work. They were deeply attached to the species and eager to understand whatever was plaguing them. They had learned in place, making their own observations of the changing forests and the changing climate, and they had learned from one another. They had sought out information on the cause. Their concerns had escalated far beyond forests and Alaska. They held a place-based sense of what Alaskans were losing right in front of them and the consequences of that loss, but they also worried about the implications of that loss on a global scale. The cypress called attention to the bigger climate problem that so many others deny or try to ignore.

I STOPPED INTERVIEWING people two towns and seven participants after Teri, when the pattern for behaviors became evident to me. I could see a relationship between the knowledge groupings and their reported actions, between certain behaviors and the ways in which individuals used and valued the forests. I wrote down a preliminary list of the behaviors people adopted:

Integrating knowledge about the dieback into public education efforts about climate change
Favoring or promoting the species in forest management
Physically avoiding the dead trees and affected forests
Creating new uses for the dead trees or affected forests
Substituting dead trees for live
Experimenting with planting yellow-cedar trees
Taking advantage of better hunting in the forests affected by the dieback
Caring for individual trees by piling snow or other activities
Limiting direct uses that negatively affect the tree population

There were other behaviors harder to attribute directly to the dieback. But some of the most knowledgeable people also reported their efforts to refrain from actions that could contribute to climate change. They were biking to work, reducing home energy use, doing whatever they could to attack the root cause of the problem. The canary called for more than adapting to changes in the local environment.

Saturation point came with some surprises. First, there was a psychological dimension to how the most knowledgeable people were coping. Some were reaching out and talking in their communities. Others imagined a forest across time and space—believing that perhaps what dies in one place could colonize another. They focused on a hopeful life elsewhere to accept death here; perhaps the trees would shift to higher elevations or higher latitudes, or just somewhere else one day. Maybe they were already doing just that. Some whom I interviewed were in mourning.

Second, even among the knowledgeable people who expressed the most fear, and who doubted whether they could do anything that could make a difference for the trees or for climate change, I found individuals still doing something. Whatever their rational conclusions about the relationship between their actions and global environmental catastrophe, some just wouldn't resign themselves to a doom-and-gloom demise.

There was the historian and fisherman in Gustavus who told me that planting a garden is an act of faith, but planting a tree or just allowing an area to grow back in trees is a supreme act of faith.

"There's no way that you're gonna be around to harvest it, or to see its fruition," he said.

There was the restoration ecologist who said the solution requires experimenting; it's worth trying to adapt and then readapt to the system that evolves along with you.

People who were attached to the yellow-cedar trees had the most at stake with climate change. But I discovered that those relationships also motivated them to act. They were forced to innovate—to hold on to what they could, to let go of what would not remain, and to find what might sustain them now. Teri told me in story. Science told me in patterns.

Everything is moving in one colossal cycle; we affect the climate, the climate affects how ecosystems function, and then the resources we depend upon become less reliable, more fleeting, maybe even disappearing completely. Ah, resources, just writing the word now makes me cringe. Every time I thought of "resources" or used it or read it after meeting Teri, the same reaction occurred in me. But it's the culture I come from. It's the language I know. If I stood up at an ecology conference to discuss the implications for forest "relationship" management instead of resource management, what would everyone think?! Fluffy! Hippie! Soft science! But the water we drink, the trees that exhale the air we breathe, the soil that feeds the plants that feed us—no ecologist denies that it's all connected. If people are going to adapt to the coming changes, the first step is accepting that we are *not* separate. So, then, what are the healthiest relationships I can create?

A manager with the Forest Service argued that there are many things we can do to help reconnect people with nature and their natural environments, wherever they live.

"You're a part of the ecosystem," she said. "Humans are part of the ecosystem. They can't be divorced from the ecosystem, but they need to know their ecosystem to understand their impact on it. It's a way of understanding."

INSTEAD OF RACING the clock to get the data before the weather window closed that spring, I raced the clock to finish interviews and reclaim my temperature sensors on the outer coast. I needed to get back to my father's home to help clean it out. We had to decide what to take and what to give away. We couldn't put those steps on hold anymore.

I returned to Gustavus again in June. Captain Zach planned to take me, Paul Hennon, and a friend from Juneau to the outer coast—out from Icy Strait, past Cape Spencer, and into Graves Harbor, where my temperature sensors were still collecting data at a subset of my plots. If they had worked properly, I'd get another year of hourly air and soil temps. They held the last, the very last, data I needed. Volunteers from an organization called Adventure Scientists in Montana went back for my sensors on Chichagof. It was a team geocache mission: enter GPS location, paddle or boat to beach, hike into forest, reclaim little button device, hike out. The task didn't require scientific training, but it did require luck, good weather, solid navigation skills, and the ability to persevere through the forested terrain.

Greg came to dinner with us again the night before we left for the coast. It was another evening at Lori's house.

"So, the people you interviewed," Greg asked. "Do they have solutions? For what we can do about the trees and what we can do about climate? What do *you* think, now?"

"I think I have a lot of coding and analysis to do before I have much to say," I replied. I didn't have the strength to talk about the experiences of loss that people had shared with me. I wasn't ready to debate with him about whether the bigger picture for us was hopeful or hopeless.

A few thoughts from the most knowledgeable people I interviewed still haunted me. I imagined Greg would say they were realistic, and I didn't want to hear that. I was tired from putting my own loss on hold, from opening myself up again and again to hear and try to understand what so many people were experiencing in relation to the standing dead.

"I think there are some things we could do about climate change, but whether we'll choose to do them, I don't think we will," one man had told me. "I think we'll look at our own pocketbook and look at the inconvenience of doing something about it and choose not to have to go through that. We'll let somebody else deal with it."

"Frankly, I think we're doomed," another said. "I really do, but I don't want to live my life with that, because if I live my life thinking we're doomed, well, I might as well go pull the trigger right now. I have a small, small glimmer of hope, but only if we greatly, greatly change. The Earth doesn't care. The Earth's gonna be here, and it'll survive. It survived much worse than us, but we're doomed. We are totally doomed as a species."

I recorded those bleak perspectives as a researcher, and I would report them, but in terms of how I wanted to live my own life—the outlook I carried every day—I just didn't want to agree.

At the door, Greg slipped on his wool button-down shirt and then turned back toward me before leaving. "I'm a bit disappointed I didn't push your buttons enough this time," he said. Was he being sarcastic? I didn't ask.

Captain Zach called me early the next morning before our scheduled departure.

"It's not looking good," he said.

"Not looking good like we can't go?" I asked.

"There's a storm moving in. Hard to tell when it'll hit exactly. We might not be able to round the point to get there. If we do, we might not be able to get back for a while. We could just wait a few days." Paul needed to return to work in town. I had a plane ticket south and my brother waiting to clear out my father's house.

"I won't be able to try again," I said. "This is all I've got."

"Well, we can run out halfway, poke our nose out, and see what it looks like."

As we put more miles between us and Gustavus, the water transitioned from glass to gentle rolling waves, then to peaks and troughs. I reached up to grab hold of a support beam on the ceiling and held tight each time the hull slammed against the sea. Far across the horizon, I could see a wall of water surging upward where Cross Sound emptied out to the coast. Lines on the surface of the water marked the fast lane, the tide stacking against the wind. Swirls and whitewater bursts.

"We'll duck into Elfin Cove," Zach said. "We can wait and see if anything changes there, but I think you're gonna need a Plan B."

Elfin Cove is a tiny fishing community built on stilts and winding docks on the northern tip of Chichagof Island. There are only a couple dozen year-round residents, but boats come in and out from the coast during the fishing season. I helped Zach secure the FV *Taurus* to the dock cleats, then sat down in the sun on the warm, wooden slats. My feet dangled over the water. Paul wandered off with the Forest Service radio in his hand.

I can't stay, I thought, *but I can't leave the sensors there.* I was stuck. I swung my rubber boots back and forth like a toddler on a big-girl chair, trying to come up with Plan B but still not willing to accept that Plan A was no longer an option.

Paul came back to the boat with word from the Forest Service Dispatch in Sitka.

"They're telling me there's a storm the size of the Gulf of Alaska brewing," he said.

"That's ridiculous," I declared, defeated. "It's dramatic enough out here. There's no need to exaggerate. How am I supposed to trust them with a report like that? A storm the size of the Gulf of Alaska?! That's really what they said?! Come on, give me wind, give me swell, give me some numbers. What a ridiculous report."

"That's what they said—the whole Gulf of Alaska," Paul shrugged. "Bad, it's bad."

We camped on a small island that night not far from Elfin Cove. The seas on the coast never calmed. I could hear the wind howling all night. I kept quiet on our ride back into Gustavus in the morning.

"You still have any grant money left for another round of fuel?" Zach asked before we reached the dock.

"Yeah, enough for gas," I said, feeling guilty about the wasted fuel for a failed mission. I was contributing to the problem in order to do research I thought would help solve the problem. What a frustrating irony. "But not enough for your time," I added.

"That GPS unit has the locations, right? Can you show me how to map each one?"

"Yeah, it's super easy," I replied. "Each plot has an ID. You just enter the ID and then there you go." I loaded one and showed him.

"And all you need to do when you get there is find these little sensors?"

"Yep, one sensor attached to a big tree with a pink flag around it. One in the soil below. There are a few different locations in both inlets."

Zach pulled back on the throttle and eased us alongside the dock. Paul jumped out to tie us up. Zach cut the engine.

"Go home," he said. "Do what you gotta do. I'll rally a few friends in town, and we'll go get them when the weather clears. Give me that GPS and get outta here."

When my father and my mother got divorced after nearly thirty-five years of marriage, my father moved to Lively, Virginia, and then north to Fredericksburg, where he bought a historic townhouse that was falling apart. He said he was more suited for small-town life and tired of the rat race. "I ain't gonna work on Maggie's Farm no more," he'd say, and then sing a few more lines of his favorite Dylan song off-key. His house was an eyesore; some of the residents nearby thought a teardown was sad but inevitable. Apparently, my father hadn't thought so. He had disappeared from our lives for about a year—maybe more, I can't remember—but communication fell off. I was concerned at first, then angry. Eventually, I had stopped wondering what filled his days, where he'd gone, what had happened. I'd finally found a way to expect nothing. Then, just as suddenly as he'd disappeared, he'd resurfaced, with invitations for my brother and me to visit.

When we finally did visit, we'd walked into a rebuilt masterpiece. He had showed us pictures of the before and after—worn floors now polished and glistening, cracked walls patched and rebuilt, fresh stucco. He and his girlfriend, Pam, had created a new home from what others had given up on. The walkway was slate. My father had chased down an ad in the newspaper and bought a ton of that slate for $50.00 from someone who just wanted to get rid of it. They'd refurbished antique furniture from cheap finds at yard sales. He passed me a photograph of people gathered around a table in the yard. Everyone was dressed in work clothes, dusty or flecked with paint, their plates piled high with food.

"What's this?" I had asked.

"The people who helped us restore this place. Once we had a vision, everything came together, bit by bit."

"And the food?" My father had always been a very talented cook, so I assumed it was his work.

He smiled. "I cooked a lot for everyone."

The pop-into-my-life / pop-out-of-my-life pattern was the rhythm of our relationship that I remember most. I don't think he ever realized how much time passed between the comings and goings. With the old, polished floorboards beneath my feet and the photograph of new friends working together, I felt a wave of forgiveness for the long disappearance. I was proud of him and what he had created. All that had happened a few years before his death.

The house was pretty much exactly how I remembered it from the first visit, but my father's girlfriend had organized a lot of his personal things before we arrived to make our task easier. My brother, Ryan, his wife, Mika, and I divvied up domains. They tackled the kitchen and garage. I took on the study, which was packed with his books, files, and music. For two days, I pored over his biographies and old photographs, records, and letters. I didn't think about climate change and the future, the sensors I'd left behind, or dead trees in the archipelago. Mika maneuvered delicately among the boxes we packed up. A bowling ball balanced on her petite frame; just two days after my father died, they'd found out she was pregnant. It was a boy.

Ryan said the relationships he held closest in life had changed during the forty-eight hours in which he lost his father and learned he would become one. Then they'd kept changing. Focusing on a new life, born amidst loss, had made his movement forward easier. I got that. He couldn't linger forever in what was, longing for a past now gone. This little one coming made that clear every day and, from what I could tell, pushed him to accept death and embrace new life at a faster rate than me. His best strategy was to adapt to the world taking new shape.

I felt lighter going back to Santa Cruz, but I carried more things. I didn't take much—the coffee mug I remembered from my childhood, some more records, and a couple of photographs of my mother and father on the beach when they were in college. The edges of the prints were tattered. I arrived home late and sat on the porch listening to the waves crash on the shore. The air was cool and damp, the fog layer thick.

A text came through from Captain Zach. He'd made it to the outer coast and back and managed to find every sensor I'd left out there. I could finally stand still for a while, stop scattering between places, and sort out how Alaskans were adapting to the changing forests while I adapted to the world without my father.

PART III: TOMORROW

The future is an infinite succession of presents,
and to live *now* as we think human beings should
live,
in defiance of all that is bad around us,
is itself a marvelous victory.

—Howard Zinn

Unaffected forests in Glacier Bay National Park and Preserve (left) and affected forests on Chichagof Island (right). Photograph by the author.

CHAPTER 10

Measured and Immeasurable

THEY KEPT THE door to the Hartley conference room locked until I was ready for everyone to filter in. It was the day of my defense. I paced up and down the narrow aisle between the rows of chairs, arranged neatly on the mottled gray carpet. Then I stood beside the podium, ran my fingers across its slick, shellacked wood, and tested the presentation slides again. Before opening the door, I bent over, placed my hands firmly on the ground, and kicked into a handstand against the front wall. My black dress slid down around my waist, and I wondered—just for a moment—if I should have double-checked the lock. But I felt strong and calm holding myself upside down, and when I stood upright again, I was ready for formality.

The room filled. My judging committee, including Paul, Eric, Kevin, and a few other renowned scientists who had advised me over the years, claimed the first row. The chairman, also a voting member of the group, welcomed us all. I took a deep breath and began.

I had buried myself in data for a year and a half to reach that day. When I hadn't been able to sit in front of a screen anymore, I'd organized stacks of transcripts across the carpet. The pages, sorted by themes and highlighted in various colors, had created a physical nest of my interviews. That nest took over my entire living space by the sea, and the door to my room became another version of the wall on the outer coast. Jump in, jump out—find my way through the forest on one side, embrace wide-open space on the other, refuel before another round.

I had ridden my bike along the coast. I'd laughed with new friends and old. I'd hiked in the mountains. I'd talked with my family on the phone, and those relationships had shifted. When I wasn't immersed in data, I strove to live as joyfully as I could. I was more aware of time—forever fleeting. I told my brother on the phone, not long before I defended my dissertation, that if I suddenly got hit by a truck, I might have some regret for all the years that had passed. I was finishing my PhD, single in my thirties, and something about the science still felt incomplete.

My colleagues and I had ended up doing exactly what I'd thought I would never do. We'd modeled future climate and projected the fate of all those healthy trees in Glacier Bay. The results were as we'd suspected. Snow cover was projected to wane, risk to turn from low to high, yellow to orange, then orange to red in the years ahead.

While I was preparing to submit my studies to academic journals, a conservation biologist from the Center for Biological Diversity contacted me about a petition under way to list yellow-cedar under the Endangered Species Act. Kiersten Lippmann, the biologist compiling the relevant information, and the others involved eagerly awaited my results from the outer coast. They wanted to know if the species was regenerating in stands affected by the dieback.

The seventy-two-page petition was filed in June 2014 with the US Fish and Wildlife Service, initiating a formal review. "Despite the survival of some individuals," it read, "once a forest becomes affected, current scientific understanding documents significant losses of this species across all life stages (from reproduction to large trees) in areas affected by the dieback at low elevations, and yellow-cedar appears to be maladapted to areas affected for the foreseeable future."[1] It was a careful way of stating what the best available science tells us about the climate-impacted areas—that it's not looking good for yellow-cedar. The petition argued that without "drastic reductions in greenhouse gas emissions and a ban on all live-logging removals, yellow-cedar will continue to suffer widespread decline." After interviewing people who earned their livelihoods in the archipelago's forests, I knew listing yellow-cedar as endangered would come at a cost. "Without cedars present, most timber sales are economically unviable," the petition acknowledged.[2] *There are always trade-offs*, I thought when I reviewed the filed document.

My ecological data revealed plants' responses to the dead and dying trees; my interviews revealed people's responses. I did, indeed, discover a relationship between knowing about the dieback and adapting to it. Knowing that climate change triggered the death of yellow-cedar trees led to different behaviors—actions targeted at the source of the problem (climate change)—not only at the local expression of it, in the forests, but more broadly. But that knowledge also came with the risk of fear, helplessness, pessimism, aversion—emotions made all the more intense by a sense of attachment to the species itself.

Thriving, for people and plants alike, required reaching toward the light. For people living the most connected to the forests, the ones acutely aware of the unwieldy effects of climate change, that meant more than just a physical response; it meant coping psychologically. It meant coping emotionally. For some, it meant refusing to live as if nothing they could do mattered. It meant refusing to live as if we are too late, already, to alter the current trajectory.

Poignant moments from all the hours of interviews I recorded played on repeat in my head as I approached the date of my doctoral defense.

A Forest Service employee I interviewed told me, "It may be the plight of the modern-day environmentalist to oscillate between utter despair and steady optimism." I'd witnessed the despair, recorded the loss, empathized with the feeling of vulnerability and helplessness. But it was optimism that inspired me; it was the people holding on to a sense of agency, who felt empowered to accept and confront what some still deny.

"You should have a 'next steps' slide," my colleague Fran said in the practice session, days before the big day.

"Next steps?"

"Yeah, right, like what you'll research next. Future work." For Fran, one of the best young scientists I knew at Stanford, it was an obvious element to include. Integrating economic and climate data in complicated models, Fran studied the rates and effectiveness of how farmers adapt their agricultural practices to climate change. That kind of research has big consequences for lots of people, because it's about understanding how climate affects what farmers are able to produce, and ultimately our global food security.

I made that "future work" slide. I practiced carefully using the word "could," as in "future work could include" studying forests affected by the dieback in British Columbia. Or it "could include" developing a large-scale survey to study—in big numbers with lots of people—how learning about the impacts of climate change on an iconic species—such as the coast redwood or the Joshua tree—could motivate action. But each time I practiced my talk in anticipation of the actual day, the next steps felt like the *shoulds* to me, like what I *should* do as a freshly minted scientist from Stanford. I *should* study another "system" where the consequences of climate change are hitting hard, where whole communities are experiencing climate change at the front lines through floods or drought or fires or eroding landscapes, where people are trying to adapt to rapidly changing conditions. *I should. I should.*

What I really wanted to do was write.

If writing is an act of discovery, and subjectivity has no place in science, I imagined that only this book could bring me resolution. What hung over me at my defense, like the last layer of fog clinging to the archipelago, was how to maintain a positive outlook amidst climatic chaos. I wanted the answer for myself, and I wanted it for us all.

Why were some people I interviewed optimistic about the future, despite all they knew about climate change?

So MUCH OF science is about systematic approaches and defensible methods; there's often little room, in discourse and time, for the lessons learned. After I presented p-values and figures showing the dynamic rearrangement of the plant communities in the forests at my defense, I had but moments to address the final question: *"So what?"*

"So what? So what's the take-home message?" I asked the audience. Ecologically, I said, despite the survival of yellow-cedars in some places, we're seeing significant losses of this species across all life stages in forests affected by the dieback. Reductions in the yellow-cedar may also create opportunities for other plant species. Which plants flourish after the death of yellow-cedar trees, when and where that occurs, all changes over time, as the years and decades pass. What thrives, ecologically, is not set in stone.

"If you just look at one point in time, you get a very different picture," I said. Timing and conditions and species traits—how they interact with each other and the surrounding environment—all matter for thriving in the wake of cedars.[3] Mosses decline early as the canopy opens; grasses increase, and different conifer species compete for positions to determine what these forests will become. Eventually, shrubs prosper in the understory. Mosses appear to return when the canopy closes again—just as the community assumes a new structure around the dead.

"Adaptation," when it comes to people, "can and does occur from the bottom up." Bottom up—meaning citizens in place changing their behaviors, and forest managers integrating what they know about climate change into whatever forests they manage and the practices they implement. All that can and does happen before anything comes from the top down in regulations and policies. (The point being, why only wait for the top to act?)

There wasn't a law restricting the use of cedars because of climate change. There wasn't a law requiring a forest manager to consider where a tree may be more likely to survive in the future, when making a decision about which trees to harvest or which to let be. But people made those choices.

"My findings suggest that when people learn from their community members about climate change impacts and interpret what they learn through the changes in their own local environment, doubt about the causes may be reduced," I said. This knowledge, along with the global concern we develop, can lead to action—like educating others, substituting other species for various uses, even reducing energy consumption. These kinds of actions can have a ripple effect; they extend beyond the local community. Environmental writer and climate activist Bill McKibben says mitigation will only work by multiplication; we have to scale up efforts to curtail emissions and change more quickly than what's comfortable—but adaptation starts in our neighborhoods.[4]

What I didn't say (because there wasn't time—nor was it appropriate in terms of scientific generalizability) was that I see the forest manager experimenting with planting trees farther north as similar to the city planner considering how high the sea will be, to the farmer in Iowa deciding which crops to grow and where, to the Montanan clearing brush to prevent forest fires from spreading. I see the logger reshaping his business to the shifting timber supply in the archipelago as similar to the Californian installing a gray-water system, the Inuit switching hunting locations, the New Yorker flood-proofing his basement, the health practitioner considering how and where climate will affect mosquito populations and the diseases they spread. These people are accepting climate change. They are integrating it into their lives. They are preparing for the future, perhaps, but they're also reshaping the present to the harsh reality of what they know. They are making it a part of their daily living and doing what's within their power to do—in an effort to adapt. All those actions begin with an understanding of a relationship with nature that climate change affects. They begin with a sense of urgency, not distance.

In a study published in 2000, David Uzzell, a psychologist studying human-environment relationships, found that an individual's perceived responsibility for the environment is greatest at the neighborhood level and decreases as the area becomes more remote.[5] When it comes to climate change, that means acceptance starts by understanding what's happening in our own communities.

Knowledge. Attachment to nature, living as a part of it, not apart. Heightened concern. Response, action. Somewhere along that journey is grief and all that comes with it. The grief I knew for my father. The grief Greg, Teri, and Ernestine knew for the trees. The grief we each had felt, myself included, for understanding the risks of continued warming—the risks for more than just the trees. I'd come to see that grieving is a necessary component of healing. From grief comes acceptance, and acceptance creates the space for restoring connections and forming new relationships. My father had taught me that. The late John Caouette had taught me that. Teri, Ernestine, Greg, and others had taught me that. This tree species had also taught me that. Not everything lost is completely gone. What fills the gap created?

I clicked beyond another slide of results and then relaxed. It was a brief pause I had planned in my presentation—a moment for connecting before proceeding to another wave of figures and projections.

"How many people here have gone hiking in a redwood forest?" I asked. The hands I expected shot up.

"Close your eyes," I told the audience. "Paint a picture in your mind of a redwood forest. The canopy above, your experience walking through it." I scanned the faces across the room. Grins spread wide as everyone lost themselves in their fantasy forests. Then I switched the slide to an image of a yellow-cedar forest—one of the fish-eye photographs I'd taken in the graveyards.

"Now open your eyes. Imagine these are the redwoods you know. How do you respond?

"Is your hiking use affected in any way? If you knew the death of these trees was related to climate change, would it motivate you to do anything differently in your daily life?"

It was the most scientific way to ask, *What will be your canary?*

"Now, trained as an interdisciplinary researcher," I continued, "I've noticed that ecologists sometimes have this 'doom-and-gloom' outlook about the future when they spend their days looking at the climate models or losses in biodiversity. Social scientists tend to be a bit more optimistic, having witnessed the ways in which human communities and institutions have evolved in response to tragedies in the past. I am some mix of the two. Following this tree for these years has allowed me to realize, more philosophically, that how people respond to a rapidly changing environment has some parallels to the ways in which we respond to other challenges in life." Sudden death. Slow death. Loss. Change forced upon us. Change we choose. New life.

I looked up from my notes and scanned the faces again. All of a sudden, I felt like I wasn't being judged anymore. Like maybe the weariness of the doom and gloom resonated, that I wasn't the only one in the room holding space for optimism amidst despair.

"WHAT WOULD YOU do if you knew what I know?"

I was thirty-one, three years shy of my PhD, when climate scientist James Hansen gave a TED talk that began with that question. I was only six years old when he said that he believed, with 99 percent confidence, that Earth was getting warmer: scientists could associate the warming with the greenhouse effect; we could expect an increase in the frequency of drought and heat-wave events.[6] Scientists criticized him for overstepping his bounds with advocacy. The public was divided between skeptics and believers, as if science were a religion we either subscribed to or didn't. Activists since have heralded him as a climate hero, calling him the Paul Revere of climate change. That question—"What would you do if you knew what I know?"—was as much about being human as it was about being a scientist.

It was June 23, 1988, when Hansen, then director of NASA's Goddard Institute for Space Studies, testified on climate change before members of the US Senate. At the time, many scientists were hemming and hawing about the magnitude of the problem and continuing to refine their models of future scenarios. But Hansen was already focusing on the urgent—the known linkages between human activity, atmospheric effects, and warming.[7] He was a physicist and a mathematician. He and his colleague Sergej Lebedeff, an atmospheric scientist, were the first to conduct a global temperature analysis using recorded surface temperatures collected over the course of a century. The results showed a clear rising trend.[8] In my eyes, if there was ever a scientist who forever altered the public conversation on climate, it was James Hansen—the canary in front of Congress.

When Hansen testified, I was too young to understand the implications of global warming. The science itself was evolving and only beginning to enter the public arena. Researchers were wrestling with the ramifications of their findings. But when I look at the photographs from Hansen's testimony today, they bring me back to the living room of my childhood home. My brother and I are sitting cross-legged on the rusty-red oriental rug beside an antique wooden chest. Sprawled out lengthwise, my father is on one couch, with his feet hanging off the far arm, and my mother, sitting upright, is on the other couch. She sips a Diet Coke and sifts through the newspaper. Hansen speaks in front of a microphone. His voice is calm, his face stern. He leans forward in his suit and tie, as if his words and figures might not be enough to reach out and grab everyone in the room.

As a little girl, I'd watched the *Challenger* explode at the Kennedy Space Center from that itchy wool rug. Sitting cross-legged in my pajamas, I'd later watched the Berlin Wall fall on the nightly news, and the protests explode at Tiananmen Square. But when I look at the photographs of Hansen in 1988, knowing what I know now, my thoughts stray to Mike Wallace from *60 Minutes* interviewing Jeffrey Wigand, the insider who blew the whistle on Big Tobacco. I was in the living room for that one, too, when it aired in 1996.

I remember expecting the truth to lead to change, and for tobacco, it did. It did for DDT, too. But even today, decades after Hansen told Congress he believed the climate change models with a confidence level of 99 percent, with better estimates, better predictions, and a better understanding of the actions that contribute to warming, with ice melting and seas rising, more lived experiences of heat waves and droughts, stress and deaths—there are still people who can't hear it, or refuse to hear it. Who say that the science is not enough. That the economic and political barriers are too great.

Hansen's global averages showed—from real, measured temperature data—what the scientific community had long anticipated. By 1861, we knew that gases could act like a blanket warming Earth's surface. Inspired by the observations and speculations of physicists Joseph Fourier and Horace-Bénédict de Saussure on the transmission of solar and terrestrial heat through the atmosphere, John Tyndall, a British physicist, discovered in a laboratory that gases, such as carbon dioxide, absorb heat.[9] Swedish scientist Svante Arrhenius constructed the first climate model in 1896; it showed that adding carbon dioxide to Earth's atmosphere would raise Earth's average temperature.[10] The greenhouse effect certainly wasn't breaking news when Hansen spoke before the Senate committee—the new, and terrifying, insight was that warming was already being documented across the planet.

In his 2008 book, *Censoring Science: Inside the Political Attack on Dr. James Hansen and the Truth About Global Warming*, Mark Bowen argues that Hansen received so much skepticism in the 1980s because no other scientist had even come close to having the insight into the global climate system that he did.[11] There is an element of intuition involved when someone understands a problem so well that the only obvious course of action is to stand up and speak out. I put Jeffrey Wigand in that pool. I put Rachel Carson in that pool, too. And Bill McKibben, when he published *The End of Nature*. They all spoke from extensive knowledge that informed their beliefs, from what they accepted as true. The birds would go silent with continued use of DDT. Smoking would continue to cause lung cancer. The gases we are still putting into the atmosphere will continue to alter the climate system. "If the waves crash up against the beach, eroding dunes and destroying homes, it is not the awesome power of Mother Nature. It is the awesome power of Mother Nature as altered by the awesome power of man," McKibben wrote when I was eight.[12]

I answered many questions in the Hartley conference room after my talk and in the two-hour closed-door session with my committee afterward. There were questions about plot design and metrics for comparing temperature regimes, questions about participant recruitment, questions about limitations, forest management recommendations, and the future of our National Parks and designated protected areas as warming continues.

It was not the science that still gnawed at me, but two other questions I needed to resolve for myself:

How do you live with what you know?

Do you have hope about the future?

These are the questions my students ask me today when they hear what I spent six years studying. These are the questions that the cab driver asks me in San Francisco, or the people sitting next to me on the plane when they peer over at my computer screen. These are the questions that stem from the bright red future projections, from a population in decline, from tables depicting what some people are losing and what they feel with the knowledge they hold, from all that occurs in the wake of the dead and dying cedars.

If you believe climate change is occurring, if you accept the inevitability of at least another degree of Celsius in warming, these questions are not alarmist; they are logical.

Nobody asked them. I couldn't have answered them fully at my defense if anyone had. They were the by-products of doing the research itself; the results of the fact that science is not removed from lived experience.

There's an objective world of the measurable—one where I can identify the species and count the saplings and run the statistics on a large dataset of forests affected by climate change. There's an objective world where the patterns from interviews reveal that the people most connected to nature are also the ones most prepared to act and respond as it changes around them. Then there's a whole realm of the immeasurable that's deeply intertwined with the measurable. Where adapting requires collaboration and working across the boundaries that climate change ignores; where mitigating the damage requires both restraint and bold action. Where what I feel is just as important as what I know. It is what comes with intuition based on knowledge—the kind of intuition that leads someone like Hansen to warn others, the kind of intuition that leads someone through the fear of what's coming to faith in our survival, from thinking "There's absolutely nothing I can do!" to "There's so much I can do!"

I NEVER USED the data from the "outlook questions" in my interview protocol for any of my analyses. These were the questions I asked at the end of each interview—about what environmental problem was of most concern and what, if anything, we could do. I never hit a saturation point for all the people who identified climate change as their top concern; there was never a moment when I could predict whether someone felt there was a lot we could do or nothing to do at all, whether the future for humanity in the face of climate change was doomed, or whether we could chart a more favorable course.

When I looked for results systematically, I also found nothing. The scientist in me said, "Well, maybe the sample size was too small. Maybe if I designed a new, large-scale survey of citizens who are knowledgeable about climate change and its effects, then I'd get a better understanding of the future outlooks they hold, how they live their days—what they do or don't do."

How do you live with what you know?

Do you have hope about the future?

I remember lying in bed the night after my defense, thinking maybe those questions lay outside the realm of science. And maybe they *should* reside outside the realm of science. If I couldn't predict who held an optimistic outlook and who embraced the doom and gloom, that could also mean we each have a choice. I have a choice.

It was up to me to decide how I want to live with the knowledge of a rapidly changing world, cut and carved and warming still. What kind of outlook do I want to cultivate? What do I want to do with what I know? I refused to wallow in any puddle of despair and helplessness; that meant optimism was my choice.

I imagined Greg Streveler would say, "Stand still. Live simply in one place, in one community. Lead by example." I had witnessed the life he'd created in Gustavus—a rooted life with minimal fossil fuel consumption. He stood still to maximize relationships in place. He rooted to take the good and the bad, to observe the changes occurring around him and to adapt, over time. Standing still, staying home, charting the changes around him, living simply, partially off the land—those are the things Greg chose to do with what he knew. *Maybe Greg's model is the ideal*, I thought. *Could I live like him?* I wondered. *No.* It didn't feel like my solution, but sometimes the most extreme responses, even the outliers, provide insight for the masses. McKibben says setting an example is vitally important, but he left his solar panels at home and got on a plane to advocate for climate action because he felt "addition alone" couldn't work.[13]

They're on two ends of the spectrum, I thought. Contributing in place and scattering for a broader reach, taking individual action and pushing for collective. I think we need both approaches and everything in between. People need to catalyze one another to create a large-scale response, like the rapid way in which the chemistry and the physics have catalyzed the warming. This isn't a situation where any one person or group has the blueprint, but we craft it together through actions big and small, through care for one another, through openness to a form of society that isn't what was or is today, but, rather, is still yet to come.

About a quarter of the Alaskans I interviewed highlighted global warming, climate change, or ocean acidification as a result of climate change as their top environmental concerns. But the answers they gave as to what to do about it were across the board:

> *"There are some things we can do. I would hope that there's more we can do if we discover even better, cheaper energy sources."*
>
> *"There are a lot of things we can do, but will we do them? Our willingness to sacrifice is limited."*
>
> *"There's a lot we can do, and we're not doing it."*
>
> *"I'm an optimist. There are a lot of things we can do. Politics and greed can get in the way of making some of the tough decisions we need to make as a country and as a world. But we're already seeing some recognize, from a social standpoint and an environmental standpoint, businesses have a role to play as well."*
>
> *"I don't think there's anything we can do."*

I'd made a spreadsheet of all those answers long before my defense. I'd searched for patterns, but all I could find was unpredictable despair and optimism, and oscillations between the two.

Large-scale surveys revealed nationwide trends, offering a portrait of America that my in-depth interviews, isolated to the archipelago, could not. Around the time I defended my dissertation in the Hartley conference room, the Yale Program on Climate Change Communication released an updated study on the public perception of climate change. *Global Warming's Six Americas* segmented the American public into six groupings based upon responses to survey questions about global warming beliefs, involvement in the issue, and other aspects of engagement. Although the researchers found one-third of the population to be doubtful, disengaged, or dismissive, that left the other two-thirds alarmed, concerned, or cautious about climate change.[14] In the March 2016 update to *Six Americas*, the numbers increased, indicating continued progress from where we were when Hansen first spoke up.

The *Six Americas* research also indicated that 45 percent of Americans belonged to the two most engaged segments of society—the alarmed and concerned. These were (and are) the people who are fully convinced of the reality and seriousness of climate change. Among them, the alarmed were already taking individual action and engaging in political and consumer activism.[15] I saw the results as uplifting, and still do, because they mean we're beginning to harness those sentiments into a powerful force. From what I witnessed in Alaska—the knowledgeable and concerned refusing to sit idle—there's more potential and motivation to harness.

A WEEK AFTER my defense, my landlord decided his family would move back into the house I was renting in Santa Cruz, so I had thirty days to find a new place to live. I took it as a sign to float free. I found a temporary spot to stay until graduation and planned to return to Alaska. I had promised the people I'd interviewed that I would return to share what I'd learned. I would present my findings from the outer coast forests and the local communities where I'd worked to the Forest Service and the National Park Service.

I edited manuscripts and made revisions to articles in review, but my attention shifted from meticulous refinements to deeper personal meaning. By resisting the pressure and opportunities to race off and study another species dieback in another spot on the planet, I created space for the meaning—my meaning—in this project. I packed boxes into another storage unit, carefully setting aside all the personal journals I'd kept from the archipelago.

When a scientist doing cancer research speaks out for human health, we call it speaking in the public interest. When an educator writes about inner-city public schools and acknowledges a shared goal for better learning outcomes, no one objects. But when a scientist speaks out about the environment, even if conditions pose a risk to people and their livelihoods—perhaps even all of humanity—we call it advocacy, or declare, "Biased!" A health-care professional can speak for the public good but the climate scientist cannot. Why? At what point will adaptation become triage, caring for the people most affected? Like an epidemic, an extreme weather event can devastate whole communities of people, but when do we start investing in getting out in front of the next one?

If I had spoken more openly at my defense, I would have said we don't have to wait or simply wish for a sweeping policy that is just as massive as the climate problem we face to tell us what to do. Waiting for the top-down approach is an excuse for doing nothing from the bottom up. Adaptation requires me to stop thinking about climate change as someone else's problem and accept it as my own. It requires me to stop thinking about the *global* risks and to start seeing what's happening in my own community, and then to reach out to others. It requires me to consider the more vulnerable places and populations to ask, "What can I do to help?" These are the things this cypress, and all the people connected to it, have taught me. What happens at the local scale matters when it comes to climate change because that's where people's lives are carried out.

IN HIS BOOK *The Hockey Stick and the Climate Wars*, Michael E. Mann, one of the scientists contributing to the Intergovernmental Panel on Climate Change Third Assessment Report, explains the science behind the "hockey stick" reconstruction of Earth's rising temperatures, which the 2001 report presented. The graph, a result of research Mann conducted with a team, had ignited a controversy. There's the blade—a relatively flat line that runs for centuries, with a little downward trend for cooling. Then there's the shaft. Enter the industrial revolution, and the line shoots upright. Mann wrote, "When we first published our hockey stick work in the late 1990s, I was of the belief that the role of a scientist was, simply put, to do science. . . . I felt that scientists should take an entirely dispassionate view when discussing matters of science—that we should do our best to divorce ourselves from all of our typically human inclinations—emotion, empathy and concern." But, he added, "Everything I have experienced since then has convinced me that my former viewpoint was misguided."[16] As Dr. Oakes, I allowed myself to stray back to those human inclinations—to emotion, to empathy, to concern.

In the six years I spent pursuing my research, I'd become a member of the most knowledgeable pool when it comes to climate change; this tree species had given me that—gift or plague, who is to say. I wanted to know what my fellow scientists who were working on climate change, and the other wicked environmental problems related to it, had experienced but kept quiet, silenced by the separation between science and the personal realm. B—behavior, in the K-A-B model—What do you do with what you know? How do you cope?

Before I went north again, I contacted a number of colleagues to open the conversation we'd never had.

I wrote:

> We all work on challenging topics that can often have a sense of doom and gloom. I don't know if this is your experience, but I often hear from people, "Where do you find hope?" when they hear about what I spent years studying. I imagine each and every one of you has heard this question before, and maybe even asked yourself the same.
>
> When you think about the future in terms of environmental change, how would you distinguish hope from faith? Do you experience either or both when thinking about the future (of planet, humanity, or self)? How is faith different from hope for you?
>
> I am welcoming any stream of conscious thoughts you are willing to share with me.

Scientists are so busy. The hours are long and to-do lists endless. Emails often go unread, or get skimmed with no reply. A colleague of mine at Stanford says she gets, on average, a thousand emails a day. I anticipated silence. But the replies came in steadily that afternoon. From the colleague working with firms to improve agricultural practices at the farm level:

> I have hope that we can create a more sustainable future (statement demonstrates my understanding that effort will be required to get there) while I do not have much faith that things will work out if we just sit back and wait.

From the colleague working with coastal communities to understand the effects of sea-level rise:

> Hope is what I want for the future given (in spite of?) what I know. Faith is what I would expect for the future based on deeply held respect for the rules of the universe: evolution, thermodynamics, relativity, etc. I hope that we (as in humanity) get our act together to control population, corporate hegemony, etc. and become better stewards of the earth that sustains us; I have faith that *eventually* things will be better; I expect (based on evidence) that they'll get a lot worse first. I don't think this is pessimism, per se, but it's one of the reasons that I have tried hard to refocus my own work on a scale small enough where I think I can make a positive difference for the local communities (people, places, systems) that matter to me personally.

From the colleague working for a land trust:

> I guess I think about hope versus faith in the same way I think about routine versus ritual. What do I mean by this? Self—I'll start here as I think this is the place I've done the most reflection. Hope and routine are similar to me; they are the daily things you do that are part of a program or a checklist. Faith and ritual are living at a higher level. You are able

> to give more to self and to others because you are focused on the important facets of life: kindness, gratitude—the fundamentals of living intentionally. Environment: I've thought a lot about the fact you must consider the rituals and faith of self before you can move on to bigger things—your work. I believe environmental challenges are somewhat caused by the gap between routine and ritual in self. I do have faith that things will work out. I am an eternal optimist, glass at least half full. I believe we are capable of change. I believe we are capable of healing. It may not be the exact, ideal, equitable solution we "hope" for, but we'll figure it out.

From the colleague working to make global supply chains more sustainable:

> For me, hope is almost like something passive. You just "hope" it's going to be okay. I resort to hope when there's nothing left that I can do. Faith, on the other hand, has a connotation of determination. It requires acting. It's not passive. I try not to think too much about the future, but in working on the environmental problems of today, and in trying to identify solutions, I am driven by faith.

I try not to think too much about the future. That one caught my attention as a way for coping. I have tried to do the same and failed completely.

We are trained as scientists to be detached—as Mann put it, to "take a dispassionate view." That's easier to do in some branches of science than others. Playful curiosity drives all kinds of research interests, from black holes to cellular function. I'd taken a modeling class at Stanford with an ecologist who had begun his career studying the flight paths of butterflies. But I believe that most of the time, interest in environmental topics originates from some level of conscious or subconscious concern. Then we end up sitting as trained scientists, debating the methods and the margins of error; all the while, there's a species facing collapse, or a community of people losing their homes to flooding, or a world turning red on the maps to indicate future warming. We (the scientists) want to get it right, and we have to get it right, and so all our energy goes into the practical, the rigorous, the absolutely critical but also mundane. My program at Stanford was an exception: we were trained to develop new insights and solutions to urgent problems—from climate change to freshwater availability to human health and sanitation. But trying not to think about the future in my own day-to-day living always felt like avoidance to me.

Esquire ran an article later that summer, after my defense, titled, "When the End of Human Civilization Is Your Day Job."[17] It was about the gloom that sets in for climate scientists and their inability to talk about it publicly. The article highlighted Jeffrey Kiehl, a senior scientist with the National Center for Atmospheric Research, who had taken a break from climate models and projections to get a psychology degree. "Ten years of research later," the article said, "he's concluded that consumption and growth have become so central to our sense of personal identity and the fear of economic loss creates such numbing anxiety, we literally cannot imagine making the necessary changes. . . . Climate scientists are different only because they have a professional excuse for detachment, and usually it's not until they get older that they admit how much it's affecting them—which is also when they tend to get more outspoken, Kiehl says." Kiehl was quoted as stating, "You reach a point where you feel—and that's the word, not *think*, *feel*—'I have to do something.'"

My work in Alaska showed me that "doing something" is not only about fighting for mitigation, educating others, or reducing home energy use through small actions; it's about finding ways to cope with what's coming. It's looking for the positive in the midst of the negative, embracing opportunities, and accepting some unavoidable losses. It's acting out of faith, getting out front in our communities and reaching across to others, not just letting hope blow in the wind. It's believing that if we got ourselves into this mess by the collective consequences of individual actions, we can also get ourselves out of it.

Faith requires acting. It's not passive. That also resonated with me. Rebecca Solnit wrote, "Hope is not a door but a sense that there might be a door at some point, some way out of the problems of the present moment even before that is found or followed."[18] But hope for me has never felt empowering. It's wistful. It feels like forsaking my own responsibility, counting on others for solutions instead of looking for what lies within my own realm of power and agency and choice—even if I am just one person caught up in a situation created by billions. Maybe when it comes to climate change, we need to leave hope behind and take responsibility for one another instead. My own optimism goes hand in hand with faith, not in the religious sense of the word, but as a conviction that we still have the strength and the power to chart a new course for the future.

I have tried hard to refocus my own work on a scale small enough where I think I can make a positive difference. I agreed with that one, too. The climate models and the global economics are important, but if another scientist, or an educator or news editor, was struggling with where to focus and asked for my opinion, I'd say the work on local impacts is more urgent today than making more global predictions. Big data has power, and the economics for mitigation requires power. But if adaptation starts in our neighborhoods, the science needs to match the scale at which individuals act and make decisions. As citizens, you and I, each one of us coping today or inevitably coping tomorrow, we need to reach a greater understanding of how climate change affects immediate relationships. Work is needed less on the red-colored globe and more on the changes occurring in local environments. What's happening in my home habitat, whether it's a city by the rising sea, a landlocked town in the sweltering heat, or a community bordering forests at risk of flames? What's happening to my source of water? What about the food I'm eating and where it grew? What will we be able to grow there tomorrow?

Two circles—Nature and Self—one on top of the other. Completely merged. That's the set I would have picked if I had turned the interview on myself. Because that's where nature is no longer an externality. Because that's where the problem is no longer us versus the other—what Naomi Klein calls the economics at war with the planetary system.[19] Because Nature and Self as merged is where responsibility begins. It comes with a consequence—the grief of loss (my research had shown me that; I lived it, too). But Nature and Self merged as one also comes with benefits—care, concern, motivation to act.

Hansen, Mann, and all the climate scientists in the *Esquire* article—their window of the rapidly changing world is framed by bleak numbers. They are restricted to a stark view. But from the window I look through, *there are a lot of things we can do.*

Wearing flip-flops in the California heat, I crammed rubber boots and dry bags into my little blue Subaru a few weeks after throwing my cap in the air at graduation. I drove north along the coast to Bellingham, left my car with a friend, and boarded the ferry. I went back to Alaska again—first as a researcher sharing results, then as a writer filling in gaps. And all the while as a citizen, carrying what I know, doing what I could do—however small my actions.

CHAPTER 11

The Greatest Opportunity

THERE'S ALWAYS SOME mix of sad and strange and beautiful in returning to a place that was once everything. It's an unexpected checkpoint for the passage of time. A record of what was. A mirror of what is. Returning illuminates what has changed and what remains the same.

I watched buildings in Bellingham fade into the distance. The ship made its way along the familiar route—from Vancouver Island, then up along the coast, and eventually past the island of Haida Gwaii. By evening the following day, we had crossed over into the archipelago. The passage narrowed. Dense forest abutted both sides. Free from deadlines in the final push to graduate, I felt the most relaxed I'd felt in years.

I set my things beside a large window and spent most of the trip writing. I took notes about hope and faith from all the emails I'd received and brainstormed questions I wanted to ask other scientists. I fixed slides for my presentation in Petersburg, the small town where the regional leadership for the Forest Service would convene. I watched the forests cruise by and waited for the first sightings of the dead and dying trees. We passed a pod of humpback whales bubble-net feeding. Fins slapping, heads popping upright, their bodies formed a circle on the surface of the sea. Air bubbles broke into rings amidst the splashing as they worked collectively below, stunning whatever school they'd encountered for prey.

What I remember most from my days in Petersburg was feeling disappointed after my presentation. Maybe my expectations were too high. I had imagined sitting in a room with the managers and decision-makers at the Forest Service in Alaska and having a real conversation about implementing new strategies for managing the population across the National Forest. I thought we'd talk more about planting, about feasibility for harvesting dead trees, about what it would take or how controversial it would be to increase protection for the trees or to facilitate their growth in more favorable habitat. These are the strategies the people I interviewed discussed. These were the actions individual managers were already experimenting with—making the best of opportunities and moderating harm in the changing climate conditions. But my talk was scheduled by the Forest Service as informational instead of decision-oriented, so that is how it went.

I was deeply disappointed. Maybe it was my anticipation. Maybe it was because I wasn't racing off to study another dieback, and wanted my research to make a difference for this one. But maybe it was yet another moment of realization that coming to any sort of consensus about what to do when it comes to climate change is just so hard. It requires knowledge and understanding. It requires belief. It requires concern. It requires listening to the science and translating it to consequences for what we, as a community or a culture or an entire species, collectively value. It requires investment—and allocating funds for risks or future likelihoods takes away from investments in today. It requires the ability to break down institutional barriers, to overcome the laws that were created before we understood climate change was even happening, and to create new ones with what we now know. Planting trees costs money and requires enormous effort. Perhaps it's a futile effort. Protecting more trees means loggers lose out.

Beth Pendleton, the regional forester, stopped me in the streets the night after my talk. Petersburg—or Little Norway—is a small town, settled by Norwegian fishermen over a century ago. I was walking along Nordic Drive, dodging fishermen and looking for a place to eat.

"Thank you for all your hard work," Beth said. "You've given us a lot to consider in terms of the future of this species."

"I want it to be useful," I replied, uplifted by the chance to talk further. "That was always my intent—to generate useful information about the ecology and the implications for Alaskans."

"Well, what those dead and dying trees mean for Alaskans is the part I find most compelling," she said. "Yellow-cedar is one piece of the system, of the forests we manage here, but it's an important one, and one deeply connected to the Alaska Native communities. So from people you interviewed, most thought the Forest Service *should* be doing something?"

"Yes," I said. "That something looks quite different for the Native weaver who wants to preserve her heritage in those trees or the logger who wants to harvest. I can't say for all Alaskans or everyone in Southeast. I only interviewed forty-five. But yes, of the people I interviewed who used and relied upon yellow-cedar, they generally felt we should be doing something—planting, protecting, favoring, collaborating across boundaries with the Park Service, monitoring. It's really a whole portfolio of shoulds."

"And the loss?" she asked. "The coping with loss is predominantly cultural?"

"Anyone attached to those trees has something to lose," I said. "But the intangible values, the ones that aren't just economical, the loss of those takes a different kind of emotional toll. It's the same challenge you've had before to balance multiple uses of the forest, but now there's the added climate stressor."

We talked for a few more minutes. She thanked me again and asked me to send the final copies of my publications.

Five minutes, I thought as we continued in different directions. *You spend five or six years studying something that becomes absolutely everything, then you get five minutes for recommendations.* I was disappointed that no one had strategized or made any decisions that afternoon, but I consoled myself with a more modest impact: the information had resonated. Beth was still thinking about it all. I had evidence that the people most connected to the trees were already adapting. I still have faith that efforts will continue.

I GAVE MORE talks in Juneau, Sitka, and Gustavus that summer. I returned to Glacier Bay as a kayaking guide for *National Geographic*. The weeks on the water gave me distance from the science, created even more space for meaning, and allowed me to share what we know about the trees and about climate with the people I led. Even in the rain and the heavy winds, I could see more beauty surrounding me without the mission of collecting data hanging over me. But those weeks also felt remarkably empty, like an escape, as if I wasn't doing enough with everything I had learned.

I vowed to get the last articles published. I vowed to keep asking questions about hope and faith and future outlook. I vowed to keep writing. I vowed to do more with what we know about climate change instead of only studying more. And I shaped everything else around those commitments: I rented a place in Juneau for a couple of weeks when the scientific journals sent me revisions to do for my articles; I took a part-time research contract back at Stanford, then a teaching one later. If I managed everything efficiently, the workload would allow me the time to keep writing.

To my friends and colleagues, I call what happened in the months that followed "the clearance window," because I could call other researchers—such as Dr. Craig Allen, the scientist who had synthesized the extent of forest die-off events around the globe—and instead of asking about species vulnerabilities to drought, or temperature increases, or statistics and methods, I could ask how his research affects his life. Whether he lives in fear. How he wakes up in the morning and moves through the world with the knowledge he holds. I did these things in the clearance window. I asked what I'd never asked as an ecologist. Working from my journals, my notes, and more interviews, I wrote during the clearance window. The kaleidoscope of colors in the California landscape cemented memories of the archipelago and all my searching amidst the shades of gray and blue and green on the outer coast.

DR. CRAIG ALLEN sent me a series of articles before we spoke on the phone. I read every one I hadn't already—the titles revealed

" . . . Predictions of Massive Conifer Mortality Due to Chronic Temperature Rise"; "Forest Ecosystem Reorganization Underway . . . "; " . . . Underestimation of Global Vulnerability to Tree Mortality and Forest Die-Off . . . "; "Larger Trees Suffer Most During Drought in Forests Worldwide."[1] Then he followed up with an article published in 2013, in *High Country News*. On the magazine cover, he wore a green United States Geological Survey fleece jacket and stood beside two other scientists dressed in plaid, flannel button-downs. They were underneath a ponderosa pine tree—its needles brown, brittle, and dead. "Tree Coroners," the article called the team.[2]

"Knowing which species survive in which conditions will help managers decide if, how, and where to try to make forests more resilient," Michelle Nijhuis wrote in the editor's note.[3]

Same for people, I thought. To make our communities more resilient, we have to decide not if, but how and where.

The photographs of ponderosa pine forests in the American West revealed the pattern I knew all too well: skeletons of dead trees across a vast expanse, standing like toothpicks on end. "Models are useful in planning for the future, but we needn't wait for them to be perfected in order to start grappling with the effects of climate change on forests," wrote the author of the "Tree Coroners" article.

Same for people. I thought of Hansen. He gave us the models so long ago.

I had questions for Craig as a colleague in the field of dead and dying trees, but I was more eager to talk about his experience doing that work. I wanted him to take me seriously as a scientist, so I felt nervous about jumping into personal questions at the outset. We needed to arrive there.

"How did you first become involved in studying forest die-off events?" I asked. It was a friendly starter that I thought would put us on common ground. Craig rewound the clock to 1989 and launched into his dissertation research on landscape changes in the Jemez Mountains of New Mexico.

Oh no, I thought. *We've got half an hour for this, maybe an hour if I'm lucky*. I was pretty sure that if we went down the dissertation road, we'd never get to the personal.

Buried in a few paragraphs of a 346-page dissertation on landscape changes in the Jemez Mountains of New Mexico was what Craig said became the most interesting discovery. The lower margin of the ponderosa pine forests had died off in the area he studied. The trees were moving upslope.[4] Craig said the observations could have been anecdotal or just a local phenomenon. But years later, when he refined the study and published it in the *Proceedings of the National Academy of Sciences*, the local went global.[5]

"You started hearing from other people who were studying similar things around the planet?"

"That's right," he said. "They provided a global change focus. It was the kind of thing people predicted could be happening elsewhere, and it was." Craig dove into the nuances of the study and the changing forests in the Jemez Mountains. I watched the time tick. I felt like I couldn't interrupt, but tried digging for the layers beyond the science.

"I guess what I'm looking for," I said, "is a more global perspective from all this work, over all these years, and your experience—"

"All right," he said. "So once we get into the 2000s, fire dominated. It was the year 2000 when the Cerro Grande fire happened. Does that name mean anything to you?"

"No, I'm afraid that we won't have time for the bigger picture about—"

"Yeah, well this is a big picture!" he interrupted. "That fire was a billion-dollar fire event."

It was only after a long explanation of the tragic event and the research following that something triggered the more personal for him. "I started considering," Craig said. "If I really think the big old trees on Earth are going to die this century, well then, what's the most important thing to be working on right now?"

"Right!" I exclaimed, relieved. The clearance window opened. "I think more in terms of a practical level for these species diebacks occurring. How do they affect people?"

"You know, if we find out that two degrees C is too much for a large fraction of the world's forests, we could be sorry about that, because we are almost committed to that now," he said. His pace shifted from the rapid delivery of numbers and dates and facts to slower, more carefully considered responses. "Yeah, there's that piece—the possible feedbacks, and the carbon cycle that will affect people too. The feedbacks are why entities like the Pentagon and many other policymakers worry about the stability of the planet from a societal standpoint. But I am talking about a profound level, ancient trees, iconic ancient old trees, everywhere I've been."

"We're finally hitting home!" I said. "Tell me more about that!"

"Well, everywhere on the planet, every society, has a deep connection with trees, particularly old ones. There are special trees, sacred trees, historic trees, memorial trees. They're embedded at every level in human cultures. But in many cases, these big old trees, the sacred, special trees on this planet, will not survive—at least not without special tending—the kinds of temperature change that we are expecting. We don't have to model to project that, as we're already seeing an awful lot of effects at one degree C. I think there is a very high probability that some of the most wonderful, special, coolest, iconic, individual, big old trees, old-growth forests, sacred groves and forests, the vegetation of this planet, will reorganize under a strikingly different climate by the end of this century. And what does that mean? There will be an increasing sense of loss among those of us who know it."

"You're *really* hitting home now!" I said. "I went into my own research thinking that people's responses would be about behavior change, like how or if they use the forests differently, whether there are signs of human adaptation occurring now. But what was most fascinating to me was the whole suite of psychological ways that people are dealing and coping with loss of an incredibly valuable iconic species that holds a long human history."

"That's exactly what I'm talking about," he said.

I hesitated to dig further, mostly because I wondered what he'd think of me as a scientist, and then stuttered out the next question: "Well, I'm just curious, but what do you think of the future? Are you afraid of future impacts?"

"I get those questions," he said. "Lately I did a bit of a lecture tour. People always ask, 'How do you work on this stuff?' 'This is depressing,' they say." He paused. "I don't see it that way. The planet is just this marvelous, beautiful place, and trees are one of the most glorious expressions of it. You have so much character when you live a long time as old growth, and ancient trees do. Just trees in general, whether young or old, there's so much we don't know about them yet. For a class of organisms that is so important to the functioning of the planet and to the well-being of human societies, it is stunning to me how little we actually know. And how little we actually invest in protecting our forests, compared to a lot of other things. So I'm motivated. It's not about being depressed, but about feeling a sense of urgency that we need to make progress in a lot of different ways, whether that's understanding these trees and forests, and how much risk they are at, or what they mean for society. Obviously, there's a strong need to change how we do energy for human societies on this planet if we want to avoid destabilizing a lot of things that people care about not only aesthetically, but also functionally. I'm energized by working from a sense of urgency."

"That strikes a chord," I said.

"In the end," Craig said, "nature is extraordinarily resilient. Life itself is so wonderfully beautiful and resilient and potent. It's like burning libraries to be eliminating species and some of the most glorious expressions of individual organisms within a species. I would consider old-growth trees and forests to be part of that. But the planet will still spin without them, and there will be forests again on this planet someday—either because humans get it together, or because we don't. Right? Old trees will grow again, but there could be a window of centuries where they're lost—where old growth doesn't exist—because the system has been destabilized. It takes hundreds of years to get old trees back again. So this is what I've been thinking about: In that window, what would be important to safeguard now so people at least know what the target was, what the stories were, and the meanings, the songs, as well as the sights and sounds and smells of our forests? You know, what is a yellow-cedar forest? I've never been in a yellow-cedar forest. What does it smell like? What does it feel like? What does that sound like? You know."

"I get what you're saying about urgency. I wake up feeling that way every day."

There's so much we can do! I thought, my new mantra. I didn't ask specifically, but it seemed safe to assume that for Craig, the canary was the ponderosa. For Craig, the canary was the fire of Cerro Grande. Over the years, he'd heard one canary after another.

I liked the idea of urgency as a positive force. If only we could all embrace the urgency and work thoughtfully from that pressing sense.

THE MORE I talked openly with colleagues, the more it seemed that every one of them, at some point in their career, had considered the questions troubling me—*What will these changes mean for me in my lifetime? And what about life beyond the blip of my own?*

I found an article a scientist named Bill Anderegg had written for *High Country News* during his time as a PhD student at Stanford. Bill had been a couple years ahead of me in the biology program when I was a grad student. He had studied Sudden Aspen Death (appropriately termed SAD), a climate-induced mortality occurring in trembling aspen trees (*Populus tremuloides*) in Colorado. Bill was a publishing machine. In the time we'd overlapped at school, I'd watched him crunch out study after study published in top journals: *Nature Climate Change*, *Global Change Biology*, *Proceedings of the National Academy of Sciences*.[6] But buried in the hard science was also a more poetic, personal piece.

"June 25, 2009. In the pale predawn light," he wrote in *High Country News*, "the branches of the aspens clawed at the sky as if trying to drag any moisture they could from the dry, cold air. . . . No one's childhood memories of a place survive returning to that place as an adult. Things seem to grow smaller, or dingier, with age, even if they haven't changed in reality. But here, the trees I remembered so clearly had changed. They were dead, all of them, their once-leafy branches bare and skeletal. I had known they would be, of course. I had returned to Colorado after six years to study the . . . aspen die-off, but I didn't foresee how visceral, how shocking it would be."[7]

When I asked him what it had been like to study a place he'd known as a child, he said it had hit him pretty hard.

"I think it was when I started to get a personal sense of what climate change was going to mean to me, that these were some of the most special places in my imagination and in my sense of self, and to come back and see them, literally, just withering away, was pretty striking. The intellectual curiosity came later, when I was reading some of the papers about the aspen die-off and realizing that it actually was really strongly connected to climate change."

"In your work today, are you able to separate the two—the scientific puzzle and the personal meaning?" I asked.

"I suspect like many, I do kind of separate the two on a day-to-day basis. You just can't live, live with the overbearing, depressing scope of things on the day-to-day. I try to focus on making steps on concrete projects and answering concrete questions with the datasets that I have. I think it's more when I step away from my work, or talk with friends, family, or journalists, that I reflect on what it means. Doing the fieldwork, experiencing what I study, brings it out, too. Whenever I'm out in the forests, I start to think about the future. The final thing I'll say is that having kids has really changed that for me as well. My daughters will hopefully still be alive by 2100, and suddenly, these climate change scenarios we keep carrying out to that year seem a lot more present when you realize that somebody you care for and love is going to be alive throughout the whole time period."

"Are there things that you do differently because of what you study?" I asked. "Has it changed your life in any way?"

Just as Craig had paused before responding to the more personal questions, Bill created the necessary space for reflection. It felt unfair of me to expect an immediate answer to something I'd struggled with for so long myself.

This doesn't have to be the definitive answer, I thought. *Your answer, my answer, Craig's answers, they will evolve again and again and again.* I waited, uncomfortable in the silence.

"I think it absolutely has," he said decisively. "My understanding of the effects of climate change is why I do what I do. I was looking for something that fit my own skills and interests in my career, but also something that would make the world a better place. I try to be as conscious as I can be about the environment in my daily activities and choices. The other thing I've realized is it has changed what I pay attention to. I see the signs around the world that have a very clear climate signal. So there's this recent mass bleaching event throughout the Pacific. What's happening with corals in this bleaching event seems to be even more widespread than what's happening with forests. There are signals everywhere."

I could see them. My colleagues could see them. What about the rest of the world? I was surprised that even at Stanford, in the class I was teaching on sustainability, my undergraduate students were only beginning to open their eyes. To listen. Early in the quarter, I'd assigned a reading from *The New Yorker* about adapting to climate change in urban planning. The author described specific events like the ones Hansen and others were predicting—a scorching heat wave in Chicago in the mid-1990s, floods in the low city centers of Singapore.[8] When I asked my students what had surprised them the most, one raised her hand and said, "This. These lines. 'Yet, even if we managed to stop increasing global carbon emissions tomorrow,'" she read, "'we would probably experience several centuries of additional warming, rising sea levels, and more frequent dangerous weather events. If our cities are to survive, we have no choice but to adapt.'"

"That surprised you?" I asked. "Tell me more."

"Well," she said. "I just didn't realize it's all so imminent. Umm, irreversible, or, more like a guaranteed trajectory, you know? I think we've been waiting, just hoping policymakers would fix the big global emissions problem."

I was surprised the trajectory was new news.

"Right, so—" I turned to the next student. "Wait, I need to rewind. How many of you didn't know about the lag times, about what's still coming even if we cut emissions?"

Three-quarters of the class raised their hands or nodded.

I was absolutely stunned. "I'm in such a bubble," I said, thinking of all my friends and colleagues who operate from that knowledge every day. "I forget not everyone knows and accepts these things."

"It just puts that much more of an onus on me, on all of us, really, to anticipate what's coming," my student added. "To do something, you know, not necessarily only to stop it, but to avoid the unavoidable consequences."

Yes! I thought. *Tell the world.*

I HAD THE chance to visit Teri Rofkar in her weaving studio, one last time, before she died of cancer in December 2016. I was in Sitka for an international conference on forest diseases, but I slipped away after the series on climate change.

Teri was waiting for me at her home. She had just returned from a one-thousand-mile journey through the archipelago on the ferry, stopping in various communities to talk about climate change.

She gave her thick ponytail a toss at the door and said, "Can you believe this? A couple years ago this was all gone. I didn't know how long I'd live, but my hair is back. Look how full it is!"

I asked timidly if I could record our conversation as we made our way up the stairs to the studio again.

"Yes," she replied instantly, seeming eager to relieve me of my own discomfort. "I trust you," she said. "My husband asked about you again this morning, and I told him, 'We have a relationship now. I trust this one. She always followed up. We've stayed in touch.'"

We sat down again beside her loom. It was empty this time.

"You're still digging deeper and working on this climate change project," she said to me. "Your commitment alone tells me something important."

"I guess, in the end, I couldn't separate myself from the science. Everything you shared about relationships—well, I do see myself, and people, as a part of nature. So if I'm studying how climate affects a species, I can't help but ask what changes in the 'environment' mean for people." I raised my hands to put "environment" in quotes.

"I'm not fearful," she said. "Is climate change gonna affect us? Hugely. Is it ultimately going to change our lifestyles? Oh yeah. Are the Earth and nature gonna go on? Mm-hmm, yes. We have our markers: look at the rise in the water; look at all that's going on. We've got our marching orders of what we need to do and not do, but we need to do it collectively. We need that level of organization, and to me, it's a level of compassion. Not only for the Earth itself, but for each other."

"Is there a difference between hope and faith for you when you think about future climate change impacts?" I asked her.

"I have faith in other people—that they will step up. But I have hope that things will get better. I think they're gonna get a whole lot worse before they get better. I guess if I felt like running around with pom-poms and saying, 'Climate change! Climate change!'—if I thought that was helpful, I think I would do it. But I think it's about getting in and doing the work and finding the nitty-gritty opportunities this is creating. It's opportunities, but people don't see it like that. It's like, 'Oh my gosh, I'm gonna have to change.' Change is what really keeps us young."

I admired her enthusiasm and optimism. She seemed unfazed by all she knew. She made it seem crazy that anyone could not care deeply about climate change; she also made it seem crazy to feel helpless.

"You weren't thinking at this scale years ago," she told me. She raised her hands beside her face, creating blinders on each side, and craned her neck forward, squinting. Then she reached out, gently touched my knee, and sat upright, smiling.

"What we have here is a catastrophic failure," she said. "What a great opportunity."

Perhaps the greatest opportunity of all.

CHAPTER 12

The Sentinels

WHILE I WAS writing about Wes Tyler and his logging operation in May 2017, I went back to Hoonah. It was the only trip I made with the sole purpose of doing book research. A few information gaps had motivated my return. I needed to fact-check background about yellow-cedar exports and logging practices during the heyday in comparison to what they are today, and I knew those kinds of inquiries would go better in person. I wanted to be able to describe Wes's operations and the town with a greater level of detail than my journal notes would allow. I went in search of facts and particulars, but closure is what that trip brought.

I picked up a set of keys at the main Forest Service office on the edge of town and walked the same narrow path to the bunkhouse. This time around, AmeriCorps volunteers nearly filled the quarters, but of all the rooms in that massive building, I ended up with the same one as before. I sat down on the cot beside my folded government-issue sheets and scanned the white-walled room from corner to corner.

This was the place where I had grieved my father, the place where I'd spent many nights listening to interviews about loss. Here I had searched for a way forward—through my father's death and through my own increasing knowledge of climate change as a looming threat. Sitting there on the same rickety cot, I remembered feeling frustrated and helpless years prior—my own emotions mirroring those of my most knowledgeable interviewees. I remembered feeling so lonely; Dylan's music and the recorded voices of people I interviewed were what kept me company. I remembered desperately seeking answers, not only for the tree and the Alaskans connected to it, but also for myself and the rest of us in the warming world, for humanity.

Wasn't it the poet Rilke who said to be patient with all that's unsolved in your heart? To try to love the questions themselves? Someday, he wrote, perhaps without noticing it, you'll gradually live into the answer.[1] I ran my cold palms across the wool blanket and suddenly realized all those emotions were gone. The uncertainty, the clinging to hope, the relentless seeking—gone.

This time, I carried a sense of calm, dare I say even grace, into that same sterile room.

I INTERVIEWED A Forest Service technician named Chris Budke about timber sales in the Hoonah district. I walked back into town to see Ernestine in her neon green home. I'd heard, via the cedar circle, that she'd survived a small plane crash the year prior. She was surprised to see me and welcomed me in. I sat by the window, my back to the ocean inlet.

"Here, could you switch with me?" she asked. "It's just with the glare, and the sunset, I can't see you. I want to see your expressions." We swapped seats.

"What do I look like to you now?" she asked. I squinted in the sunlight streaming through the glass.

"A dark silhouette," I said. She laughed. The house felt warm after walking in the cold spring air; the carpet, soft beneath my bare feet.

"How's your niece? Is she still here?"

"Catherine? Yes. We got a mountain goat this year. We worked on the hide all winter. We've finally gotten the fur free, and we're about ready to weave it with bark. But my hands, ever since the accident, well, I'm not ready." She massaged her right forearm with her left hand. I told her I was writing, still writing. I noted the colors in the room and the drawing on the wall of the eagle and raven again. She reached for a rose made of yellow-cedar bark on a shelf and handed it to me.

"For you."

I stopped by the same tiny shop for groceries and skipped over the twenty-five-pound bags of rice and cane sugar on my way to macaroni and cheese. As I was leaving, a man entered the store. He paused as we passed, then stared for a moment, but I kept on my way. Halfway across the street, something drew me back to him.

"I've seen you before," he said. I'd thought the same but couldn't place him. I looked down at his hands. Strong and calloused, they gave him away.

"Yes! You're the carver—Owen. I was here four years ago, and I interviewed Gordon Greenwald while you were carving." I remembered his jet-black hair, pulled back into a ponytail—now salt and pepper.

"Can you believe it?" I said, feeling embarrassed. "I'm still working on the same thing." He took my hand in his and shook it gently.

"Well, we're loading up those totems on Monday if you're here and want to lend a hand," he said, referring to the trees they'd been carving when we'd first met.

"I've been researching and writing all these years, and you've been carving?"

"Some things take time. Creating takes time," he said. "You're right on schedule," he laughed. "We just finished. We're shipping them on the ferry to Glacier Bay. They'll be installed at Bartlett Cove, where our people came from long ago."

"I have to leave on Sunday," I said, stunned I'd miss this full-circle moment by a day. With my teaching schedule, I had only a few days in Hoonah.

"Do you want to come see the carving shed again?"

We walked down the street to the warehouse where I'd interviewed Gordon. Two massive poles, carefully wrapped in blue tarps, rested on a trailer bed out front.

The sweet cedar aroma slipped out into the open when Owen opened the door. Gordon cocked his head, trying to place me for a moment, then exclaimed, "Hah, you're back," and beckoned me in. "You saw the totems are ready to go?!"

"Talk about timing!" I replied.

"We've been finishing some paddles made from yellow-cedar today. I'm working on the design for a new totem, thinking about the story it will tell." Hand-carved paddles, glistening wet with lacquer, hung from a rack across the shop.

Gordon crossed his arms and scratched his white hair for a moment, still shocked to see me. We talked for a while about what I'd learned in the years that had passed. I told him about the outer coast study—what I'd discovered in terms of the poor yellow-cedar regeneration and the shift to western hemlock–dominated forests. I talked about the hunters I'd interviewed who were taking advantage of the deer feasting on the shrubs and blueberry brush that flourished in the forests affected by the decline. I shared stories of the managers I'd met who were experimenting with planting yellow-cedar farther north along the coast—where the climate could be more suitable in the future.

"There were other Native people I interviewed, like you," I said, "who didn't know about the scientific research on the cause of the dying trees, but proposed climate change based on their own observations." On the recorder, in our interview earlier, Gordon had told me it was his "gut feeling" that something about our changing climate was affecting the species.

"We're always aware," he told me again. "We're always observing what's around us. What do you see? What's happening? Is there something I need to think about here and do differently?" I had read and listened to his interview so many times that it was as if we were reliving the first conversation.

"I was always raised with that question: What do you see? A lot of the elders, Native elders, that I worked with and was around, they always would say, 'Be observant. Be aware. It may save your life.' No matter how subtle it may seem, it may be the important thing you need to remember."

"The people most connected to nature may be best positioned to see what's changing and adapt," I said, picking up where we'd left off. They nodded.

"Have you heard about the year of two winters? Do you know Wayne Howell?" Gordon asked.

"I know him. Greg introduced us a few years back." Wayne was an archaeologist in Gustavus. I knew he and Greg were working on a study about the timing of the ice advance in Glacier Bay during the Little Ice Age, but I had limited information about the project. "The year of two winters? Never heard anything about it. What do you mean?"

"1754," Gordon said. "There's a story, passed down by an elder in Hoonah, about a strange year, long ago—a year with two winters. Around the time the berries came out, the spring turned to harsh winter again with snow falling. It kept falling. Wayne told some other scientists, and they went to core a bunch of trees. It was like using Western science to test oral tradition. You know what they found?"

"The year of two winters?"

"Yes. 1754–1755. People doubted the story, thought it was some Native myth. But scientists found the story in the trees. The trees held the same record as our people. Talk to Wayne. It's not really my story to tell, but that year was around the time the ice had reached its peak, and it tells us about the timing of when Hoonah was first settled."[2]

"And about our ability to adapt to a changing environment," I added. They nodded, in sync. Owen said the year of two winters had also helped pinpoint when the ice pushed the Tlingit out of Glacier Bay—the exodus—probably a couple years prior.

"What are you doing tomorrow?" Owen asked.

"Going to talk with Wes Tyler. He's picking me up at the bunkhouse, and we'll go out to the mill."

"Do you want to come paddle with us in the morning?"

"What do you mean?"

"Paddle with the tribe. We're taking out the dugout canoes. You could come with us, and we'll drop you across the waters at Wes's mill."

"What time?" I asked.

"8 a.m., by the canoes out front."

I hadn't anticipated a morning on the frigid waters in my packing for the short trip. I figured I would just wear absolutely everything I brought with me—however ridiculous I looked.

"I'll be there," I said.

"Tell Wes we are bringing you in style." He smiled.

ABOUT A DOZEN community members gathered around the canoes in the morning. Owen walked alongside the hull of his boat, pacing its full length with his hand on the gunwale.

"Forty feet long," he said. "Spruce. One giant spruce." Sealed to protect the wood from the waters, it was painted bright red, inside and out, with black accents on the bow and the stern.

"We hollowed it out by hand, and then heated rocks. With the wood wet inside, the canoe stretches and arcs around the scalding heat of the rocks. You can see the ends tip upward and the width expanded at the center. That's all from our work. That's all from our hot rocks."

He divided us into two groups of paddlers, and I introduced myself to the people in my boat. I didn't know if Owen had told them anything about me, and I probably gave the impression of a tourist popping through. But when I mentioned I was a researcher and writer and what I studied, everyone relaxed and fired off questions about the trees on the outer coast. They wanted to know about the health and status of their sacred cypress.

"All hands on!" Owen directed. We leaned forward on the boat, shoving the trailer wheels into motion.

We pushed the canoes, one after another, through the streets of Hoonah toward the water. A few people came out of their homes to join us, adding more hands and heart to the mission.

"Hoo haa!" Owen called out.

"Hoo haa! Hoo haa! Hoo haa," the tribe called back—the same chant I had heard from Ernestine beside the loom in her home.

We launched the boats straight into the sunlit water, loaded up, and shoved off with the butts of yellow-cedar paddles against the rocky shore. I sat in front of Owen, who steered from the stern. A whale surfaced in the distance and then its calf followed. I looked back at him, and he nodded his head to look toward the mountains.

"Shahg-wah," he said, drawing out the soft a's of a word or words I didn't know. He spelled it quickly in Tlingit, but I couldn't catch all the letters over the sound of the water rushing along the hull. "The name of the peak there on the left." (Later, I learned there are various conventions in circulation for spelling the Tlingit language, but "Shaagu.áa" was one currently accepted version of the name.[3])

I fell into the familiar rhythm of paddling, thinking about the effort and foresight it must have taken to move a whole community when the ice came over 250 years ago. To find the tree trunks on islands or in inlets that were large enough to carry families and provisions; to cut those trees and carve those canoes and shape them by fired stones, to locate new food sources, to paddle in search of whatever was next in the ice and cold and the penetrating wet. I thought about the determination required to build new lives in different climate conditions than the ones where those lives had begun. It took knowledge of what was coming, yes, but far more than just that. It took staring at the oncoming disaster straight on and not looking away. It took tremendous effort and collaboration and careful planning to resolve. It took leaders and innovators and collective action. It took faith. The faith we need again today.

"Hoo haa!" Owen hollered again. The paddlers in his daughter's boat sang in reply. We cut across in front of the village and a few more people emerged from their homes to watch our boats, *Eagle* and *Raven*, soar by.

"Hoo haa!" a man on the front dock started another wave of call-and-response.

"No pictures," a woman said from the bow of our canoe. I had my phone out for taking notes but slipped it back into my pocket. She raised her paddle high into the air and all the others joined her. Arms extended by the reach of yellow-cedar touched the sky together.

"Hoo haa! Hoo haa! Hooooooo Haa!"

I DEFINITELY ARRIVED in style, and soon thereafter, Wes's truck came barreling down the dirt road, kicking up a cloud of dust. He gave me a big, warm grin as the paddlers pushed out to sea; apologized for the mess of spark plugs, chains, and grease in the truck; and began answering questions about timber sales and logging exports and grades of wood with admirable patience. I wasn't sure how much time we'd get, so I kept my questions flowing at a fast clip. The piles of yellow-cedar, spruce, and hemlock logs lying out on the property had grown taller in the four years that had passed. Stacks of lumber and paneling, ready for shipment, filled the warehouse.

Somewhere along the tour, he stopped at a yellow-cedar log, straight but twisted with striations like a stick of buttery taffy. "This is dead standing. Here's what I'd do, if I could find a customer. I'd work it over. I'd clean it up. I'd make a centerpiece in a house, a post, or a mantel. But you gotta do some creative thinking, and you gotta have some creative inquiries."

BETWEEN MEETINGS WITH my students back at Stanford, I thumbed through a publication that Chris Budke, the forest technician in Hoonah, had given me about logging on the Tongass. I kept thinking about the year of two winters. It had nothing to do with my ecological studies on the outer coast. It wasn't even really about yellow-cedar; they'd used hemlock and spruce trees in the study. But it was about climate. It was about the strength of people, whole communities, in the face of frightening change. There was something optimistic embedded in that story. About accepting change. About letting go. About taking action to adapt—to find a way to thrive in seemingly uninhabitable climate conditions. And what had brought me there, to this year of two winters, was all my years of chasing the cypress and the people connected to it, trying to find answers in science, insight from conversations, and meaning in the act of writing.

I sent an email to Wayne Howell, the National Park Service archaeologist in Gustavus, and set up a time to talk. Wayne said he'd worked with two Gregs—Streveler, of course, and Greg Wiles, an expert in glacial geology and tree rings at The College of Wooster in Ohio.

"The winter following winter was something I'd gotten into just sitting and visiting with elders in Hoonah," Wayne told me on the phone.

"I think I remember from our meeting years ago that some of your research showed not only where the ice was but also when people had started using yellow-cedar trees in the new places they settled," I said. "You mentioned a grove on Pleasant Island. Can the trees tell a story of what the climate was and where people went? Scientifically?"

"We'd been banging around Pleasant Island for years," he answered. "I stumbled upon some small patches of yellow-cedars when I was out deer hunting on the north side of the island, but—well, that was fifteen years ago now. Several of us decided to hike across the island, point to point. We spent three days backpacking across the middle of the island that nobody ever goes into. Lo and behold, we were walking along the first day, and stumbled upon this grove of enormous cedar trees, old and really big cedar trees. We didn't spend any time with them then, other than to marvel at them. The circumference of one was probably thirteen or fourteen feet, you know, reaching—"

"Wait. Whoa! Really?"

"—reaching around and then just using my arm span as a way to calibrate the girth of it," he continued. "It was big. We went back again and looked around more. I started finding a lot of indicators of bark harvest—some of which had metal tool marks and some of which had, at first glance, no marks. Then, as I spent more time with them, I actually found evidence of stone tool marks."

Wayne had told Greg Wiles about it. Wiles had gone there and cored a bunch of the trees, taking samples both within the scars and outside of the scars to estimate the timing of the bark stripping.[4] "They got a sequence of dates that went back to 1675," Wayne told me.

"It's really interesting because it's an isolated patch," he added. "I mean, I'd tramped around Pleasant Island a lot, and all of a sudden we came across a patch of cedars where these folks have been going for hundreds of years, three hundred years of continuity. The Little Ice Age advance and retreat spans that time period; so it was an important resource patch for them throughout the entire time period, and through all the transitions they went through—ecologically and socially. It continued to be an important resource for them."

Relationship, I thought. *It continued to be an important relationship for them in a changing climate.*

The way Wayne summarized the oral tradition, when the ice came and the people left Glacier Bay, they separated into three groups. One group went to the northeast side of Chichagof Island, but stayed only a year or two. The conditions were harsh; cold winds swept off the glacier across the waters. They picked up camp and moved on, and within a year or so found the site sheltered from the north wind—Hoonah.

"It was during the founding of Hoonah that the cold snap happened—the year of two winters," he said.

I was well aware of the differences between the causes of the two crises: cooling back then because of natural variability, and warming today because of human activity. But there was something about the human response during the ice advance that felt uplifting to me. The trees at Pleasant Island survived the ice coming and going; they held a record of people moving throughout the area as the climate shifted.

Mary Beth Moss, the Park Service anthropologist in Hoonah, said she had collected numerous oral history recordings from elders about the year of two winters.

"Where Wayne and I differ," she told me when we talked, "is on the timing for when the people left Glacier Bay. Wayne backs into the date of the exodus by considering two sources of information: first, the year of the two winters as revealed in the tree-ring record, and second, oral history from several elders that suggests some time passed—perhaps several years—between the exodus and the settling in Hoonah when the year of two winters occurred. Some people interpret oral history differently than others. I've said repeatedly, 'I am sorry, but I personally don't believe anybody is gonna sit there and wait when a glacier's ten feet away or even one hundred feet away or even one thousand feet away from you.' I think people left much earlier than when the glacier actually hit the villages."

"That's a really good point," I replied.

But I wasn't thinking about expanding ice plowing over villages in a cool period of historical climate. I was thinking about sea-level rise flooding cities, crops in drought, ghostly graveyards of dying trees on every plant-covered continent. I was thinking about the climate change humans are causing, about the inevitable warming and the lag times between what we've already emitted into the atmosphere and the conditions headed our way as a result.

We're not just going to sit here and wait. I refuse to believe that. I refuse to allow that.

"I mean, is that not logical to you?" she asked. "It would have been freezing cold. I think it's a mistake to say the exodus occurred in 1752. It's possible their houses stayed there, but they probably moved out sometime before or began their movement long before then."

"Or seen it coming," I added, "and thought of other places they might go, or how they might need to change their way of life."

"For what that's worth," Mary Beth continued, "it seems entirely illogical to me to think that people would have remained at a site that was, for all intents and purposes, an icebox. It's illogical to me." She shared more information from an elder about a couple of village sites the people tried before settling in Hoonah. The place they attempted first wasn't suitable—too cold and harsh and exposed—so they picked up and tried again.

"I don't know if this is meaningful to you or not," she said.

"It totally is," I replied. "I agree with you. It seems entirely illogical that they would sit and wait and watch this happen. We're seeing an entirely different type of climate effect now—the one we have created and are still creating. But it's the same message for me. You're not sitting up until the last minute waiting for crisis to hit or for the glacier to go over the village. You're thinking about where you might go. You're thinking about what you can do."

It wasn't just illogical to me. It was incomprehensible, inconceivable, really. Waiting for the ice to hit, like waiting for home to go up in flames.

There's a huge difference between an ice face and an atmosphere, and we don't have a Planet B. But not getting out front, not doing everything we can to see what's coming and to try to adapt to it—all that is inconceivable. I don't think it's a question of what's the one best thing to do with so many bests possible. *There is just so much to do!*

INTERMITTENT CORRESPONDENCE WITH Greg Streveler continued in the clearance window. About a year after my graduation, he wrote to me: "You've made me think a lot about that species: why I love it, how it fits into the world, what makes it so vulnerable (and therefore especially precious?), why we treat it (and everything else) as we do."

"How do you live a joyful life with the knowledge you hold of the impacts coming and the trajectory that we're on?" I asked him when I saw him last in Gustavus. It was the late summer of 2017 and I was up north to teach a field course for Stanford sophomores in Sitka. We were meeting at the Sunnyside market, near Four Corners. With a few tables and chairs out back, it was a comfortable spot for coffee and conversation.

"One piece of it is intellectual. You just can't get up in the morning and lead a coherent life out of darkness. But in terms of how I actually generate joy—I don't always succeed at generating it, but it starts with being glad to be alive, glad for all the beauty around me, and then exercising that beauty. Finding ways to actually participate in it."

"That can be a simple thing, like how you greet someone on the street, how you give your attention to a plant, or a scientific question," I replied. "That's exercising beauty for you?"

"Or how I look into a beautiful mind," Greg said.

"I would also argue the beauty radiates," I mused. "Maybe the impact is on the small scale of individual lives, but maybe it's also on the scale of that larger trajectory of this planet. How can innovating, and mitigating, and taking the kinds of actions Naomi Klein puts forward for economic and political restructuring all be a part of exercising that beauty, not just about fighting against something or hoping for something else?" (Klein has argued that our economic system as we know it must be transformed in order to avoid the worst impacts from climate change.[5])

"I was, at one point, feeling rather badly that the lifestyle I lead seems to be more of an anachronism," Greg said. "I don't see any tendency toward the society heading anywhere near the direction I chose. Then one day I woke up and said, 'Streveler, you damn fool! What makes you so hubristic that you think you should lead in a way that other people will follow? You should lead in a way that you follow. And whatever value it has to others will be up to them to decide.' I no longer maintain the conceit that my life or my thoughts are exemplary. They simply are the best expression of who I am, and whatever value that has to the world, I will allow it to have. That's how I see it."

"So, it's not just what *we* can do, it's also what *I* can do," I said, trying to grasp Greg's meaning. "And maybe that pushes someone else to reflect on their own choices and ask, 'Can I meet you there?' If not, what still applies?"

"We're all human," Greg said. "There are things that are generalizable. But you have to start from the specific. In my case, one thing that makes me more and more unusual in society is my absolutely ferocious and irrevocable sense of place. One of the principles I've been able to articulate to myself is the Hippocratic principle: first, do no harm. For me that means holding still so that my life can be examined in a minimally kaleidoscopic fashion. So that I can look at a carrot I raised and I know where the nutrients came from. I know where the defense of that carrot from pathogens came from. I know pretty much precisely what I asked of the ecosystem to feed me that carrot." Greg looked down at the grass beneath his feet, seemingly in a moment of appreciation for the complex web of life.

"If I go to the grocery store in Juneau," he continued, "I can't do that. So, for me, the first thing I want to know is 'How do I avoid harm?' I have to hold still to do that, just given who I am. But most people, to them, their life doesn't start from that principle. They don't have to hold still. As a matter of fact, it's stultifying to them to hold still. So, I can't say to somebody else, 'Hold still and your life will make sense.' And there have been times in my life when I have piously pontificated on that. 'Why don't you guys hold still?' And everybody looks at me like I'm from Mars."

"What would you say this tree has taught *you* over all these years?" I asked.

"I'll reframe the question a little bit and ask, 'Why do I find this tree so beautiful?'" He scratched his beard, shuffled his feet underneath his seat, and leaned toward me.

"Let me take you to a grove you haven't seen, a little one I heard about not far from here. I wanted to go alone the first time I saw it. There it stood on this little bench above a muskeg—probably forty trees, ranging in age from some fifty to two hundred years, I'd guess." He crossed his hand in his lap. "Walking up to them was like walking up to a Rosetta Stone. Their rarity on that spot was arresting. The beauty. The first thing I noticed was the wind in their branches. It was a nice day, so I could catch the smell. I walked inside that grove, and it was almost like I walked inside a conclave of monks. It was like something I felt when I was a kid and the Trappist monks would come to our church to hold a high mass."

Greg's eyes welled up with tears, but none fell. "I could hear them singing. For some reason those groves do that for me differently than any other tree does. For some reason, for me, that is what those trees evoke. Certain things speak to us differently in this world. Maybe that's the bond you and I have—that this tree speaks to us both."

I agreed and disagreed at once. I admired his reverence, but the tree's beauty wasn't what caught my attention anymore. Nor was it the call of the canary. That tree was my looking glass into the rapidly changing world. It taught me about the tangible effects of climate change when the future projections seemed too distant and removed. It taught me how what happens to my environment will ultimately affect me. It gave me fear. It pushed me through hopelessness. But it also taught me to believe that what I do matters in spite of how seemingly insignificant I am in the face of climate change. It taught me to believe the future isn't on fire everywhere, that I can be a part of a larger force working from a place of care and concern and conviction.

As a scientist, I think the best I can do is contribute to better understanding of what climate change means at the local scale, for that's where people will need to know the risks and benefits, where lives are carried out—not just people's lives but all life—where adaptation can only come from mitigating losses (wherever possible) and embracing opportunities (wherever possible). Just like the forester, the farmer needs to know what crop to plant, when, and where. The city planner needs to know where the sea will be. The forest manager is looking for the places where the trees will flourish, while others fade away.

As a citizen, I can embrace some level of restraint in my own actions that contribute to the problem. I can reject the politics that attempt to silence the science, to bury what we know. I can help increase awareness, instead of enabling denial. I can push for relationships (not resources), and for reclaiming all we have lost through the misguided belief that environmental health is external to our own. For me, all this starts with an acceptance that I am not immune to the havoc, and that climate change knows no boundaries. It starts at home in my garden, where the care I put in determines the care it offers. It starts in the classroom, on a plane, in the taxicab, or at a dinner party—whenever the "double-c word" surfaces. It starts with working from a greater sense of responsibility for the disparities this problem creates—and exacerbates. It starts by asking, "What relationships do we want to sustain?" Then, "How and where can I help us do just that by integrating what we know about climate change?"

"I think for me, this tree has become an emblem of optimism," I said to Greg. I couldn't relate fully to what he felt amidst those trees—the parallels to his childhood experiences in church. But my faith is in our will for survival, in our intellectual capacity to problem-solve, and in our responsibility to one another. The little actions alone can't stop climate change, but they will be a part of adapting to it.

"Those trees on Pleasant Island thrived during the Little Ice Age," I said. "Many of them are still thriving now. It's a beautiful species, yes, but it's also this symbol that a species can persist, the way it has."

Greg furrowed his eyebrows. "It's not that for me. It's a symbol of vulnerability."

"This all began that way for me—knowing the species has a vulnerability to climate that we all do. That there's a threshold we may hit, too. What is that threshold?" I asked. "But now, when I come across a tree standing tall with green foliage sweeping down, I see that tree differently. It's like, 'Here you are. Despite all odds, here you are. Still.'"

I WENT TO see the sentinels of Pleasant Island, the trees that had witnessed the ice coming and going, the trees still towering with wispy, green limbs today. Wayne Howell guided me there. We pushed through thick blueberry brush with our hands held out and up to protect our faces. We stumbled over the humps and avoided the holes between roots and downed trees. We trekked across thick layers of moss with careful footing and balanced on the cushiony ground.

In an hour and a half of trekking across the moss and the muskegs, I never noticed any sign of the cypress, not even a sprig of foliage or that sweet smell in the wind. Then I skirted around a hemlock tree and there they were—yellow-cedars in all shapes and sizes, saplings and giants, gathered in small groups across the landscape.

"There, let's go there," I said, pointing toward the biggest crown I could see in the distance.

I walked inside a cluster of cedars and stood beside the two largest individuals at the center. The others formed a ring around me.

So these are the survivors, I thought. *Luck, by where they happen to be.*

Perhaps those trees had possessed some evolutionary advantage over all the others from the beginning. From seedling to sapling to ancient elder, perhaps where they had first rooted had somehow sealed their fate for survival. The ice never hit them. Protected from root injury, they somehow persist today. Perhaps their survival was, indeed, only luck. I will never know.

In the end, I think hoping is like wishing for luck, and people are much smarter than that. I'm unwilling to accept the idea that we will all just stand still, stick to the status quo, wait, watch the temperatures rise, and see what happens to the relationships we need.

What does this tree have to teach us? From 2010 to 2017, for nearly eight years of my life, I wrote that question over and over and over again. In my notebooks. In various computer files. On scraps of paper. On butcher paper hanging on my wall. *What does this tree have to teach us? What does this tree have to teach me?*

My answer:

That we are all vulnerable. There may be survivors, carrying out their lives in pockets where conditions remain favorable. They may regroup, perhaps even evolve, and at the right moment in time even flourish again.

But what does this tree ask of me? Perhaps that one is far more important.

My answer:

That I can observe the changes occurring around me and embrace the struggle to accept them, to respond to them, to adapt to them. I can look ahead and live today holding space for tomorrow. I can fight for what we can still curtail. I can play a part, not live apart, and I can act with care for others when the floods hit, when the seas rise, when the snow melts, the rivers run dry, and the flames rage. Defeat may only be a failure to adapt.

If fear is the absence of breath, and faith is a positive force, I want to breathe into an uncertain future. If this tree species and all the people connected to it gave me one great gift, it is this: the realization that there's simply no imaginable tomorrow—no modeled future scenario, no amount or shade of red—that could ever possibly nullify the need for unwavering care and thoughtful action today. To me, that is thriving. To me, in this rapidly changing world, that is grace. It is how I choose to live with what I know.

A yellow-cedar tree recently affected by the dieback. It retains its primary and secondary branches, but the foliage has fallen.

Epilogue

IN THE YEARS since I finished my research on the outer coast, other scientists have answered more questions about the yellow-cedar. John Krapek, a master's student at the University of Alaska in Southeast Alaska, and Dr. Brian Buma, a forest ecologist and colleague, conducted a two-part study in a series of groves that were recently discovered off the road system near Juneau. The first part investigated whether there was any real pattern on the landscape as to where yellow-cedar occurs. The answer was no.

"It's completely random," Brian told me. There wasn't a combination of variables—like slope, elevation, solar radiation, or wind exposure—that could predict where to find the species throughout its range.[1] The fanciest stats and finest fieldwork confirmed what many people I'd interviewed had observed—it is a mysterious species. We still don't know why it's found in some places and not in others.

The second part examined whether yellow-cedar was colonizing new areas in the region.[2] It was the question of migration—if they're dying in one place, are they moving to another? The trees they studied inside the healthy grove were all similar in age and size. Brian and John never located any younger trees outside the edges of the groves. Brian said it was as if the perfect set of conditions allowed those yellow-cedar trees to establish at the right moment in time, and now they were standing like punctuated ants, waiting for another right time to march forward.

No one knows for sure when or if that moment will come again.

The US Department of the Interior published a notice of findings and initial status reviews on the petition to list the species under the Endangered Species Act in the *Federal Register* on April 10, 2015. The petition had produced enough information to warrant consideration of a listing, so the review process continued. It made national news. The *Los Angeles Times* reported that "the Alaska yellow-cedar edged one step closer to being listed as a threatened or endangered species after the U.S. Fish and Wildlife Service announced that the tree may warrant such protection because of the ravages of climate change. The move was applauded by environmentalists while a timber industry trade group called it 'pretty silly.'"[3] An arduous and controversial review has since ensued.

In October 2017, researchers and forest managers from state and federal agencies, private consulting firms, nongovernmental organizations, and academic institutions convened in Juneau to discuss the best available science on yellow-cedar. I was invited along with other experts from Alaska, British Columbia, and farther south. Supported, in part, by the US Fish and Wildlife Service, the meeting's goal was to inform the listing review process.

Paul Hennon picked me up at the airport and greeted me as Dr. Oakes. We drank beers, ate salmon burgers, and talked about the latest maps of the dieback the night before the meetings. He'd retired from the United States Forest Service after publishing what I deemed his magnum opus in 2016. It was a 382-page report on pretty much everything known scientifically about yellow-cedar, its past, its present, and its future.[4] He and his wife, Susan, had spent their first year of retirement in Portland, but I could see that his heart and his scientific curiosity had stayed with the trees in Alaska.

Nearly all the scientists I'd worked with over the years were present in the room the next morning. To me, it felt like coming home.

"We're here to inform our Species Status Assessment, which is a science-based document that looks at the biology of the species," began Steve Brockmann, the lead biologist from the US Fish and Wildlife Service. He was warm and welcoming, informal in attire, formal in approach. Thirty experts sitting in front of their notes and computer screens formed a U at the desks around him.

"I want to establish a few ground rules," he said. "We're not looking for any sort of group recommendations or consensus. You're here to help us understand the ecology, the species needs, and the outlook. The sessions are organized that way. What you provide us in your PowerPoint presentations will become federal records—all part of the federal administrative record on this. So we will consider all that in our report of the species status. We're not recording, so what you say isn't part of that record. But we're listening. We're taking it all into account, and we'll follow up, too."

I tend to rely heavily on compelling graphics and photographs instead of overwhelming text in any presentation. But under the new terms, I opened my PowerPoint and began adding all the numbers I had only planned to say.

"What does this species need to survive?" he asked. "What are the current conditions? How are those needs being met or not met? What are the future scenarios? Both the pessimistic and the optimistic? We've got a great agenda. So let's begin."

Paul kicked it all off. Others followed at a steady clip. We moved from the historical ecology to the current species genetics and ongoing debates about botanical names. Then we got into the real pressing stuff—where yellow-cedar thrives, where it dies, where it lives on. Scientists presented maps of the dieback from aerial surveys and recent assessments of the status across the Canadian border. Brian reported results from a study we'd published showing that half the areas in the northern reaches of the species range with a currently suitable climate for yellow-cedar are expected to warm beyond that threshold by the late twenty-first century.[5] Someone presented results on successful seedling plantings. Another slide showed saplings dying from the roots up and recent observations of the dieback in young-growth forests. Analyses from forest inventories indicated that the dieback had stalled in some parts of the archipelago; I showed maps and evidence of it spreading in others.[6] Farther south, beyond the island of Haida Gwaii and other parts of British Columbia, the species appeared unaffected.

Brian called the phenomenon a "transitional mortality"—death centers around specific climatic zones, but life persists beyond.[7]

I asked about modeling methods. I asked about regeneration. I asked Steve about the review timeline.

"They call it a twelve-month finding," he told the group, "but it always takes longer—primarily because of a national backlog of petitions waiting in line for the limited funding available to address them."

"Your job is gonna be super interesting," I said to one of his colleagues taking notes at the meeting. "But also, not easy."

"That's how it goes," he said. "You've either got a population in total collapse or a complicated situation. In this one, we've got logging. We've got climate. We've got a species extending across country borders. Not easy."

No one asked us about our opinions. Like Steve said, it wasn't about voting or any consensus; it was about science. And if you were to ask me on the street or in a coffee shop today, *Should the government list the species under the Endangered Species Act?* I would probably dance around the question there as well. I'd say it's my job to share what we know, "the best available science," and to acknowledge there are also many things we still don't know. It's how I am trained to contribute objectively. We all did just that in Juneau. But, oh my gosh, even writing that makes me cringe. As Greg Streveler says, scientists have proven, again and again, a remarkable capacity for simply monitoring a species to extinction. We observe. We report. We let others decide.

Wearing my Dr. Oakes hat, I'd give you the stats on regeneration in the affected areas on Chichagof, as I did in that room. I'd offer you the future projections for the dieback continuing to spread across the region and acknowledge some debates in the research community about the more recent spread of the decline.[8] I'd tell you that migration will likely be too slow, if at all; that the rates for the tree moving elsewhere are unlikely to keep up with the rapid rate of climate change. Then I'd remind you that in the forests where I worked on the outer coast, a small portion of the population is still alive in areas affected by the dieback. They are not all standing dead. For some reason that no scientist has explained yet, there are yellow-cedar trees that continue living in a world so different from the one where they first rooted. They are not *all* sold for logging—in fact, relatively few trees go to board feet on the timber market today, and the vast majority are not even accessible for harvesting without more roads. I'd also say that, if you're interested in giving the species its best shot, then the logical course of action is not to harvest the live ones in places where they're more likely to survive. Perhaps the dead can offer a substitute. If you're interested in maintaining uses of them, that requires a careful balancing act between the values that come only by taking from them and others that require their continued existence.

A few of us agreed—yellow-cedar is shifting from a fairly widespread species to one that will be far more restricted. Even if it doesn't go extinct in the near future, any plan for harvest needs to take account of where there's potential to do significant damage.

I said good-bye to Paul at his home in Juneau on my way back to California. He was out back in the pouring rain, wearing navy blue rubber field gear and tending to his overgrown garden. We hugged beside a few yellow-cedars—seedlings, now small trees, he had planted years ago, and then sustained by piling snow at their bases over all the winters since. Paul said he was unsure which home, Portland or Juneau, would become home in the years to come.

I wish I'd said so at the time, but I'm sure he will never leave these forests completely. On the plane headed south to my partner, Matt, and our home together, I watched the archipelago slip away beneath the cloud layer again. Tears slid down my cheeks. I felt a pulse of sadness, wondering if I would ever see the yellow-cedar trees on the outer coast again. But mostly what I felt was overwhelming gratitude for what I carry with me through whatever happens next.

Acknowledgments

Weeks before I first traveled to the outer coast in 2011, I received a call from Stacey Woolsey (now Stacey Wayne), a resident of Sitka and a Stanford graduate I'd never met.

"You have quite a challenging endeavor in front of you," she said. "I want to help." So many more people shared this sentiment in the following years. My field research would not have come together without the community it created, and the same holds true for this book.

In Search of the Canary Tree evolved out of a short series I wrote for the *New York Times* while I was working on the outer coast. I thank Nancy Keeney, Justin Gillis, and Sandra Keenan for the opportunity to contribute to the Green Blog. That series, along with a narrative-science course I took with Tom Hayden and Lucy Odling-Smee at Stanford, convinced me I had much more to write. I would like to thank Tom for encouraging me from the start and mentoring me through all that unfolded.

Hank Lentfer and I never left the conversation we started on the crane flats in Gustavus in 2012. Together we logged so many hours in explorations of grief, resiliency, and connection to nature. His edits made this book better. His thoughtful questions made me think more deeply and express myself more freely, and he grounded me in my own experience whenever the intellectual took over. Our two-person "Write On!" book group sustained me.

They've never met, but in my little writing world, Emily Polk was "Hank's Other Half"—always pushing me for descriptive prose and vulnerability. "Put us there." "Tell me how you felt." I am grateful to Emily for reminding me that an afternoon spent fixing the wind in words—getting everything just right—is both a beautiful privilege and a tremendous responsibility.

Rob Jackson, Richard Nevle, Shannon Swanson Switzer, Emma Hutchinson, and Kim Kenny read the chapters month by month and provided such thoughtful feedback. Our somewhat secret Stanford writing group, which Emily first convened, kept me going and producing on deadline. Just writing and breathing in the same room with Russ Carpenter helped as we fueled each other's endeavors one Tuesday at a time.

There are far more "characters" to this story than I could name and give the proper space they deserve. Megan Barnhart and Robin Mulvey doubled our power for measuring plots in 2011. Gregg Treinish of Adventure Scientists and Corey Radis joined me and Tomas Ward in 2012 for another data-collection sprint on Chichagof. Along with pilot Avery Gast, Buddy Ferguson at Ward Air was instrumental for aerial surveys. Mark Kaelke, Phil Mooney, and Peggy Marcus provided bear safety training and bestowed the skills needed to coexist with brown bears while taking thousands of measurements. Thanks to Tom Koos at Stanford for his remote communications and first aid assistance, and to the Sitka Ranger District Dispatch for our radio check-ins. Adam Andis and Scott Harris provided kayak safety training. Lewis Sharman provided logistical support in Glacier Bay National Park and Preserve, and I thank the National Park Service for authorizing my research permits. Solan Jensen, Aleria Jensen, Kevin Hood, Julie Scheurer, Julie Bednarski, Paul Barnes, Melissa Senac, Matt Davidson, Bess Ranger, Molly Kemp, Anissa Berry-Frick, Clay Frick, Ed Neal, Zach Stenson, Lisa Busch, Davey Lubin, Art Bloom, and Chris Lunsford opened their homes and offered their cars to me between field stints and whenever I returned to Alaska to share my work or fill gaps for this book. Captains Zach Stenson, Paul Johnson, Charlie Clark, and Scott Harris provided transportation to the outer coast and assistance for my boat surveys.

Melinda Lamb, Kitty LaBounty, and Thomas Hanley assisted me with their expertise in local botany and methods for understory data collection. Mark Riley, Karen McCoy, Frances Biles, Richard Carstensen, Jonathan Felis, and Dustin Wittwer were helpful in geographic information system (GIS) mapping, working with remotely sensed imagery, and resolving GPS data-collection methods. Ashley Steel's statistical expertise contributed to what became the chronosequence.

I'm grateful for my colleagues and mentors in the Emmett Interdisciplinary Program in Environment and Resources (E-IPER) for always seeing the value of interdisciplinary research despite its challenges, and for those in the Department of Biology who welcomed me to their disciplinary world: Naupaka Zimmerman, Bill Anderegg, Holly Moeller, Maria Del Mar Sobral Vernal, and Rachel Vannette. Robert Heilmayr, Rachelle Gould, Fran Moore, Amanda Cravens, and members of the Dirzo, Lambin, and Ardoin research groups provided much feedback on my work between 2009 and 2015. Thanks to the family of Stanford's School of Earth, Energy, and Environmental Sciences, the Woods Institute for the Environment, the Department of Earth System Science, and E-IPER, especially Pam Matson, Peter Vitousek, Helen Doyle, Danielle Nelson, Jennifer Mason, and Deb Wojcik. The E-IPER staff provided critical guidance for navigating university policies and interdisciplinary research. Jennifer Mason and Siegrid Munda at Stanford, and Roxanne Park at the Forest Service helped manage the chaos of the funds. Advising from their fields of expertise, Eric Lambin, Nicole Ardoin, Rodolfo Dirzo, Kevin O'Hara, and Paul Hennon helped me develop the skills I needed to analyze the relationships between changing social and ecological conditions.

Other research assistants from Stanford, UC Berkeley, and Palo Alto High School helped during my doctoral work and afterward while I was writing this book: Ramona Malczynski, Caitlin "Captain Sitka" Woolsey, Nikhil Junnarkar, Amanda McNary, Morgan McCluskey, Julie Scrivner, Emily DeMarco, Diniana Piekutowski, Emma Fowler, and Michaela Elias. I am grateful to them for volunteering their time.

Many scientists, forest managers, economists, historians, lawyers, and other experts helped fact-check various sections of this book and provided additional information. I would like to thank Paul Hennon,

Sarah Bisbing, Brian Buma, Nicole Ardoin, Mike Osbourne, Hari Mix, Rob Dunbar, Andrés Baresch, Tom Waldo, Jim Mackovjak, Inga Petaisto, Sawa Francis, Connie Adams Johnson, Chien-Lu Ping, Katey Walter Anthony, Gustaf Hugelius, Colin Shanley, John Krapek, Madonna Moss, James Simard, Wayne Howell, Mary Beth Moss, and Buck Lindekugel. Walks and play dates with Alexis Hawks refreshed me; she regularly exchanged music with me for samples of my latest prose and jumped to make the trek north to find the grove at Pleasant Island. Mary Cantrell helped with editing references and checking sources.

I wrote most of this book while teaching in the Program in Writing and Rhetoric (PWR) at Stanford University and working part-time for Dr. Nicole Ardoin. I learned a great deal about crafting narrative by surrounding myself with scholars who talk about writing and research writing in the field of rhetoric (which I didn't know existed until I entered it). I thank PWR for allowing me to embed myself in its community and to share my love for storytelling in environmental sciences with the young, eager students who took my classes. The Environmental Reportage Residency at the Banff Centre for the Arts and Creativity gave me the space and community to develop the structure of *In Search of the Canary Tree* early in my writing process. I thank Colette Derworiz, Andrew Reeves, Maddie Gressel, and Michelle Nijhuis for the insights they offered into reporting; and Erin Johnson at Island Press, Tom Hayden, and Curtis Gillespie for their comments early on.

Many places where I wrote seeped a sense of place into this book. From the west side of Santa Cruz to the redwoods in the Marin Headlands to the intersection of Gault and Darwin Streets, coastal California lit memories of the Southeast seas. Thanks to Noa Lincoln and Dana Shapiro for welcoming me to their farm on the Big Island; to Alexandre Muller in Le Levancher, France, for letting me work in the mountains in a country where climate change is accepted like gravity, and whether we should do anything about it isn't a question up for debate. My days with Hank, Anya, and Linnea on Lemesurier Island brought me back to the shades of blue, green, and gray when I needed them most.

Jessica Papin, my literary agent, stood by me from the moment my pitch packet reached her inbox through everything that followed. This book became real when she joined my team. Thanks to Liz Carlisle, Zac Unger, Juli Berwald, Wade Davis, Brooke Williams, Terry Tempest Williams, Dan Fagin, and David Quammen for what they each shared with me about their approaches to writing. They helped me discover my own process. Liz Carlisle and Susan Hennon came sweeping into the final sprint with careful eyes and valuable input. TJ Kelleher and Leah Stecher at Basic Books contributed such thoughtful comments and edits. Their feedback helped me soften the scientific language, expose more of myself, and rewrite the beginning to how it always needed to be.

The scientific research I conducted while pursuing my degree at Stanford was financially supported by a number of grants and fellowships from the following sources: the National Science Foundation; the Emmett Interdisciplinary Program in Environment and Resources and the School of Earth, Energy, and Environmental Sciences, Stanford University; the Morrison Institute for Population and Resource Studies, Stanford University; the Haas Center for Public Service, Stanford University; the Wilderness Society; the George W. Wright Society; the National Forest Foundation; and the USDA Forest Service (Forest Health Protection and the Pacific Northwest Research Station in Juneau, Alaska). Cascade Designs, Ibex, and Patagonia sponsored me during my field endeavor by providing affordable and reliable gear for the tough conditions.

So, where is my crew from the outer coast today?

Kate "Maddog" Cahill designs timber harvest plans, burns slash piles, marks timber, and tries to keep loggers in line—what she calls "forestry at its dirtiest" and what she always wanted to do. She was always determined to become a Registered Professional Forester (RPF) in California, which requires at least seven years of forestry experience. After graduating from UC Berkeley, she left West Oakland for a small town on the edge of the Sierra and worked under the giant sequoias as a tree coroner during the California drought before moving to Ukiah in Mendocino County.

P-Fisch earned his master's in forestry and is now a forester in western Washington. He works on conservation-oriented projects for public, community, and private forest managers.

Tomas Ward now lives in Wisconsin and calls our summers in the archipelago the culmination of the scientific fieldwork he did in his twenties. He tends to the architectural heritage of his community through remodeling and historic restoration work and serves as a land steward on his family's farm.

Odin Miller is finishing his graduate studies in Fairbanks. He worked for the Alaska Department of Fish and Game, traveled to southern Siberia to volunteer as a research assistant for a friend working with reindeer herders, and then returned to school. When we were last in touch, Odin told me his experience on the outer coast shifted his perspective, making climate change more tangible, more personal, and more immediate than ever before. He used to feel that it didn't make sense for anyone to preach about climate change as long as they had any polluting habits themselves, that people should lead by example. But in recent years, he had become involved in activism by joining a movement focused on building sustainable economies in Alaska's communities, elevating the voices of Alaska Native people living at the front lines of climate change, and keeping fossil fuels in the ground. "I still feel anxious or pessimistic about it sometimes," he wrote me, "but at least I feel I'm doing the best I can to help create an alternative to the business-as-usual future. I think the solutions are going to come from building a popular movement that engages the masses and shows a positive way forward."

I never could have conducted my ecological fieldwork without them.

Lastly, family—I thank my mom, Pat, my brother, Ryan, and sister-in-law, Mika, for their steady encouragement. I thank my father, George M. Oakes (1945–2013), for teaching me to seek meaning and understanding in everything, to speak my truth, and to stray from any well-traveled path with a curious heart and mind. To Matt, who read Hemingway out loud and engaged in many musings on what matters most; who let me paint that room yellow and claim it as my writing lair; who slipped his hand away from mine only to nudge me before dawn and say, softly, "Go write," in the tradition of all the other writers in his family—thank you for supporting the ending of what began before you, too, showed up and changed everything.

Notes

NB: I upheld my typical standard of scholarship by annotating the manuscript with references and noting other relevant sources for specific topics. For the sake of keeping the in-text references thorough but not exhaustive (or too distracting from the narrative), some sources are not provided for references to uncontroversial and general facts that can be confirmed with a key word search. I did not include an exhaustive list of all the people I interviewed because some opted to maintain their anonymity in the approval process, and it was difficult to track others down again in the remote communities where I worked. Those identified in the text offered their consent.

Many of the findings I share from my own research can be found in three scientific studies that were published in peer-reviewed journals:

Oakes, Lauren E., Paul E. Hennon, Kevin L. O'Hara, and Rodolfo Dirzo. "Long-Term Vegetation Changes in a Temperate Forest Impacted by Climate Change." *Ecosphere* 5, no. 10 (2014): 1–28.

Oakes, Lauren E., Paul E. Hennon, Nicole M. Ardoin, David V. D'Amore, Akida J. Ferguson, E. Ashley Steel, Dustin T. Wittwer, and Eric F. Lambin. "Conservation in a Social-Ecological System Experiencing Climate-Induced Tree Mortality." *Biological Conservation* 192 (2015): 276–285.

Oakes, Lauren E., Nicole M. Ardoin, and Eric F. Lambin. "'I Know, Therefore I Adapt?' Complexities of Individual Adaptation to Climate-Induced Forest Dieback in Alaska." *Ecology and Society* 21, no. 2 (2016).

PROLOGUE

1. Don David and Aylmer Bourke Lambert, *Description of the Genus Pinus, Illustrated with Figures; Directions Relative to the Cultivation, and Remarks on the Uses of Several Species: Also Descriptions of Many Other Trees of the Family Coniferae* (London: Messrs. Weddell, Prospect Row, Waldorth, 1824).

2. Dominick A. DellaSala, Paul Alaback, Toby Spribille, Henrik von Wehrden, and Richard S. Nauman, "Just What Are Temperate and Boreal Rainforests?," in *Temperate and Boreal Rainforests of the World: Ecology and Conservation*, edited by D. DellaSala (Washington, DC: Island Press, 2011), 1–41.

INTRODUCTION

1. Aldo Leopold, *A Sand County Almanac and Sketches Here and There* (New York: Oxford University Press, 1949).

2. Justin Gillis, "Climate Chaos, Across the Map," *New York Times*, December 30, 2015, https://www.nytimes.com/2015/12/31/science/climate-chaos-across-the-map; Chelsea Harvey, "Greenland Lost a Staggering 1 Trillion Tons of Ice in Just Four Years," *Washington Post*, July 19, 2016, https://www.washingtonpost.com/news/energy-environment/wp/2016/07/19/greenland-lost-a-trillion-tons-of-ice-in-just-four-years/?utm_term=.846b862a2b22; John Upton, "Oceans Getting Hotter Than Anyone Realized," *Climate Central*, October 5, 2014, www.climatecentral.org/news/oceans-getting-hotter-than-anybody-realized-18139; Emily Atkin, "Climate Change Is Killing Us Right Now," *New Republic*, July 20, 2017, https://newrepublic.com/article/143899/climate-change-killing-us-right-now.

3. T. R. Allnutt, A. C. Newton, A. Lara, A. Premoli, J. J. Armesto, R. Vergara, and M. Gardner, "Genetic Variation in *Fitzroya cupressoides* (alerce), a Threatened South American Conifer," *Molecular Ecology* 8, no. 6 (1999): 975–987.

4. John Muir, "Timber Resources of Alaska," *Pacific Rural Press* 18, no. 19 (1779): 8.

5. Charles Sheldon, *The Wilderness of the North Pacific Islands: A Hunter's Experience While Searching for Wapiti, Bears, and Caribou on the Larger Islands of British Columbia and Alaska* (New York: Scribner's Sons, 1912).

6. Paul E. Hennon and Charles G. Shaw III, "Did Climatic Warming Trigger the Onset and Development of Yellow-Cedar Decline in Southeast Alaska," *Forest Pathology* 24, nos. 6–7 (1994): 399–418.

7. Patrick C. Taylor, Ming Cai, Aixue Hu, Jerry Meehl, Warren Washington, and Guang J. Zhang, "A Decomposition of Feedback Contributions to Polar Warming Amplification," *Climate AMS* 23, no. 18 (2013): 7023–7043, doi:10.1175/JCLI-D-12-00696.1.

8. J. M. Stafford, G. Wendler, and J. Curtis, "Temperature and Precipitation of Alaska: 50 Year Trend Analysis," *Theoretical and Applied Climatology* 67, nos. 1–2 (2000): 33–44; T. F. Stocker, D. Qin, G. K. Plattner, M. Tignor, S. K. Allen, J. Boschung, A. Nauels, Y. Xia, V. Bex, and P. M. Midgley, eds., *Climate Change 2013: The Physical Science Basis*, Working Group I Contribution to the Fifth Assessment Report of the Intergovernmental Panel on Climate Change (Cambridge: Cambridge University Press, 2013).

9. Lauren E. Oakes, Paul E. Hennon, Kevin L. O'Hara, and Rodolfo Dirzo, "Long-Term Vegetation Changes in a Temperate Forest Impacted by Climate Change," *Ecosphere* 5, no. 10 (2014): 1–28, doi:10.1890/ES14-00225.1.

10. Lauren E. Oakes, Paul E. Hennon, Nicole M. Ardoin, David V. D'Amore, Akida J. Ferguson, E. Ashley Steel, Dustin T. Wittwer, and Eric F. Lambin, "Conservation in a Social-Ecological System Experiencing Climate-Induced Tree Mortality," *Biological Conservation* 192 (2015): 276–285; Lauren E. Oakes, Nicole M. Ardoin, and Eric F. Lambin, "'I Know, Therefore I Adapt'? Complexities of Individual Adaptation to Climate-Induced Forest Dieback in Alaska," *Ecology and Society* 21, no. 2 (2016): 40, doi:10.5751/ES-08464-210240.

11. Wallace Stegner, *Collected Stories of Wallace Stegner* (New York: Random House, 1990).

12. David Wallace-Wells, "The Uninhabitable Earth," *New York Magazine*, July 9, 2017. This is the most read *New York Magazine* article to date. The original article sparked an enormous amount of controversy given its dark outlook, which was based upon extensive reporting and recent scientific research. The annotated version provides a wealth of references and more detailed information from the scientific literature.

CHAPTER 1: GHOSTS AND GRAVEYARDS

1. John P. Caouette, Marc G. Kramer, and Gregory J. Nowacki, "Deconstructing the Timber Volume Paradigm in Management of the Tongass National Forest," US Department of Agriculture, Forest Service, Pacific Northwest Research Station, PNW-GTR-482, March 2000.

2. Paul E. Hennon, Carol M. McKenzie, David V. D'Amore, Dustin T. Wittwer, Robin L. Mulvey, Melinda S. Lamb, Frances E. Biles, and Rich C. Cronn, "A Climate Adaptation Strategy for Conservation and Management of Yellow-Cedar in Alaska," US Department of Agriculture, Forest Service, Pacific Northwest Research Station, PNW-GTR-917, January 2016, 382.

3. Erin L. Kellogg, ed., *Coastal Temperate Rain Forests: Ecological Characteristics, Status and Distribution Worldwide* (Portland, OR: Ecotrust and Conservation International, 1992).

4. Elizabeth Bluemink, "Warming Trends: Trees Fall to Climate Change," *Juneau Empire*, March 26, 2006, http://juneauempire.com/stories/032606/sta_20060326002.shtml.

5. Paul E. Hennon, David V. D'Amore, Paul G. Schaberg, Dustin T. Wittwer, and Colin S. Shanley, "Shifting Climate, Altered Niche, and a Dynamic Conservation Strategy for Yellow-Cedar in the North Pacific Coastal Rainforest," *BioScience* 62, no. 2 (2012): 147–158.

6. Paul G. Schaberg, Paul E. Hennon, David V. D'Amore, Gary J. Hawley, and Catherine H. Borer, "Seasonal Differences in Freezing Tolerance of Yellow-Cedar and Western Hemlock Trees at a Plot Affected by Yellow-Cedar Decline," *Canadian Journal of Forest Research* 35, no. 8 (2005): 2065–2070; Paul G. Schaberg, Paul E. Hennon, David V. D'Amore, and Gary J. Hawley, "Influence of Simulated Snow Cover on the Colder Tolerance and Freezing Injury of Yellow-Cedar Seedlings," *Global Change Biology* 14 (2008): 1–12; Paul G. Schaberg, David V. D'Amore, Paul E. Hennon, Joshua M. Halman, and Gary J. Hawley, "Do Limited Cold Tolerance and Shallow Depth of Roots Contribute to Yellow-Cedar Decline?" *Forest Ecology and Management* 262, no. 12 (2011): 2142–2150.

7. Colin M. Beier, Scott E. Sink, Paul E. Hennon, David V. D'Amore, and Glenn P. Juday, "Twentieth-Century Warming and the Dendroclimatology of Declining Yellow-Cedar Forests in Southeastern Alaska," *Canadian Journal of Forest Research* 38, no. 6 (2008): 1319–1334.

8. Melanie A. Harsch, Philip E. Hulme, Matt S. McGlone, and Richard P. Duncan, "Are Treelines Advancing? A Global Meta-Analysis of Treeline Response to Climate Warming," *Ecology Letters* 12, no. 10 (2009): 1040–1049.

CHAPTER 2: STAND STILL

1. Alaska State Legislature, Final Commission Report: Alaska Climate Impact Assessment Commission, March 17, 2008.

2. C. Tarnocai, G. Canadell, E. A. G. Schuur, P. Kuhry, G. Mazhitova, and S. Zimov, "Soil Organic Carbon Pools in the Northern Circumpolar Permafrost Region," *Global Biogeochemical Cycles* 23, no. GB2023 (2009); G. Hugelius, J. Strauss, S. Zubrzycki, J. W. Harden, E. A. G. Schuur, C.-L. Ping, L. Schirrmeister, et al., "Estimated Stocks of Circumpolar Permafrost Carbon with Quantified Uncertainty Ranges and Identified Data Gaps," *Biogeosciences* 11 (2014): 6573–6593.

3. Henry C. Cowles, "The Ecological Relations of the Vegetation on the Sand Dunes of Lake Michigan," *Botanical Gazette* 27, no. 3 (1899): 95–117, 167–202, 281–308, 361–391; Chadwick D. Oliver and Bruce C. Larson, *Forest Stand Dynamics*, updated ed. (New York: John Wiley and Sons, 1996).

4. Kevin L. O'Hara, *Multiaged Silviculture: Managing for Complex Stand Structures* (Oxford: Oxford University Press, 2014).

5. Lawrence R. Walker, David A. Wardle, Richard D. Bardgett, and Bruce D. Clarkson, "The Use of Chronosequences in Studies of Ecological Succession and Soil Development," *Journal of Ecology* 98, no. 4 (2010): 725–736.

6. Roman J. Motyka, Christopher F. Larsen, Jeffrey T. Freymueller, and Keith A. Echelmeyer, "Post Little Ice Age Glacial Rebound in Glacier Bay National Park and Surrounding Areas," *Alaska Park Science* 6, no. 1 (2007): 36–41.

CHAPTER 3: FEAR AND FORESTS IN A CHANGING CLIMATE

1. Henry David Thoreau, *The Journal, 1837–1861*, edited by Damion Searls (New York: New York Review Books, 2009).

2. Abraham J. Miller-Rushing and Richard B. Primack, "Global Warming and Flowering Times in Thoreau's Concord: A Community Perspective," *Ecology* 89, no. 2 (2008): 332–341.

3. Charles G. Willis, Brad Ruhfel, Richard B. Primack, Abraham J. Miller-Rushing, and Charles C. Davis, "Phylogenetic Patterns of Species Loss in Thoreau's Woods Are Driven by Climate Change," *Proceedings of the National Academy of Sciences* 105, no. 44 (2008): 17029–17033.

4. Michelle Nijhuis, "Teaming Up with Thoreau," *Smithsonian*, October 2007.

5. Craig D. Allen, Alison K. Macalady, Haroun Chenchouni, Dominique Bachelet, Nate McDowell, Michel Vennetier, Thomas Kitzberger, et al., "A Global Overview of Drought and Heat-Induced Tree Mortality Reveals Emerging Climate Change Risks for Forests," *Forest Ecology and Management* 259, no. 4 (2010): 660–684.

6. James A. Johnstone and Todd E. Dawson, "Climatic Context and Ecological Implications of Summer Fog Decline in the Coast Redwood Region," *Proceedings of the National Academy of Sciences* 107, no. 10 (2010): 4533–4538.

7. Edward Abbey, *Beyond the Wall: Essays from the Outside* (New York: Holt, Rinehart and Winston, 1984), xvi.

8. Paul E. Hennon, Charles G. Shaw III, and Everett M. Hansen, "Dating Decline and Mortality of *Chamaecyparis nootkatensis* in Southeast Alaska," *Forest Science* 36, no. 3 (1990): 502–515; Amanda B. Stan, Thoman B. Maertens, Lori. D. Daniels, and Stefan Zeglen, "Reconstructing Population Dynamics of Yellow-Cedar in Declining Stands: Baseline Information from Tree Rings," *Tree-Ring Research* 67, no. 1 (2011): 13–25.

9. Martha Martin, *O Rugged Land of Gold* (New York: Macmillan, 1953).

10. Alwyn H. Gentry, "Patterns of Neotropical Plant Species Diversity," In *Evolutionary Biology*, edited by M. K. Hecht, B. Wallace, and G. T. Prance (Boston: Springer, 1982).

CHAPTER 4: SOLVING PUZZLES

1. Paul E. Hennon, Charles G. Shaw III, and Everett M. Hansen, "Symptoms and Fungal Associations of Declining *Chamaecyparis nootkatensis* in Southeast Alaska," *Plant Disease* 74, no. 4 (1990): 267–373.

2. Jim Pojar and Andy MacKinnon, *Plants of the Pacific Northwest Coast* (Vancouver, BC: Lone Pine, 1994).

CHAPTER 5: COUNTDOWN

1. François de Liocourt, "De l'aménagement des sapinières," *Bulletin Trimestriel, Société Forestière de Franche-Comté et Belfort, Julliet* (1898): 396–409.

2. Gary Kerr, "The Management of Silver Fir Forests: De Liocourt (1898) Revisited," *Forestry* 87, no. 1 (2014): 29–38.

CHAPTER 6: THRIVE

1. Michael Q. Patton, *Qualitative Research and Evaluation Methods*, 3rd ed. (Thousand Oaks, CA: Sage, 2002).

2. Sandra Díaz and Marcelo Cabido, "Vive la Différence: Plant Functional Diversity Matters to Ecosystem Processes," *Trends in Ecology and Evolution* 16, no. 11 (2001): 646–655; Sandra Díaz and Marcelo Cabido, "Plant Functional Types and Ecosystem Function in Relation to Global Change," *Journal of Vegetation Science* (1997): 463–474.

3. Charles Darwin, *On the Origin of Species* (London: John Murray, 1859).

4. Anja Kollmuss and Julian Agyeman, "Mind the Gap: Why Do People Act Environmentally and What Are the Barriers to Pro-Environmental Behavior?" *Environmental Education Research* 8, no. 3 (2002): 239–260.

5. Martin L. Parry, Osvaldo F. Canziani, Jean P. Palutikof, Paul J. van der Linden, and Clair E. Hanson, eds., *Climate Change 2007: Impacts, Adaptation and Vulnerability*, Working Group II Contribution to the Fourth Assessment Report of the Intergovernmental Panel on Climate Change (Cambridge: Cambridge University Press, 2007).

6. V. R. Barros, C. B. Field, D. J. Dokken, M. D. Mostrandrea, K. J. Mach, T. E. Bilir, M. Chatterjee, et al., eds., *Climate Change 2014: Impacts, Adaptation and Vulnerability. Part B. Regional Aspects*, Working Group II Contribution to the Fifth Assessment Report of the Intergovernmental Panel on Climate Change (Cambridge: Cambridge University Press, 2014). The

Fifth Assessment Report provides an updated definition from the Fourth Assessment, "reflecting progress in science" (p. 1758).

7. On learning about an environmental impact indirectly or by experiencing it directly, see D. W. Rajecki, *Attitudes: Themes and Advances* (Sunderland, MA: Sinauer, 1982); Matthias Finger, "From Knowledge to Action? Exploring the Relationships Between Environmental Experiences, Learning, and Behavior," *Journal of Social Issues* 50, no. 3 (1994): 141–160; Heidi L. Ballard and Jill M. Belsky, "Participatory Action Research and Environmental Learning: Implications for Resilient Forests and Communities," *Environmental Education Research* 16, nos. 5–6 (2010): 611–627; Brenda A. Fonseca and Michelene T. H. Chi, "Instruction Based on Self-Explanation," in *Handbook of Research on Learning and Instruction*, edited by Richard E. Mayer and Patricia A. Alexander (New York: Routledge, 2011), 296–321. On whether people felt concern, see P. Wesley Schultz, "New Environmental Theories: Empathizing with Nature. The Effects of Perspective Taking on Concern for Environmental Issues," *Journal of Social Issues* 56, no. 3 (2000): 391–406; P. Wesley Schultz, "The Structure of Environmental Concern: Concern for Self, Other People, and the Biosphere," *Journal of Environmental Psychology* 21, no. 4 (2001): 327–339; P. Wesley Schultz, "Inclusion with Nature: The Psychology of Human-Nature Relations," in *Psychology of Sustainable Development*, edited by Peter Schmuck and P. Wesley Schultz (Boston, MA: Springer, 2002), 61–78; P. Wesley Schultz, Chris Shriver, Jennifer J. Tabanico, and Azar M. Khazian, "Implicit Connections with Nature," *Journal of Environmental Psychology* 24, no. 1 (2004): 31–42. On whether the issue was one they felt they could address, see Albert Bandura, "Self-Efficacy: Toward a Unifying Theory of Behavioral Change," *Psychological Review* 84, no. 2 (1977): 191. Finally, on whether people had developed an attachment to the place impacted, see Jerry J. Vaske and Katherine C. Kobrin, "Place Attachment and Environmentally Responsible Behavior," *Journal of Environmental Education* 32, no. 4 (2001): 16–21; Nicole M. Ardoin, "Toward an Interdisciplinary Understanding of Place: Lessons for Environmental Education," *Canadian Journal of Environmental Education* 11, no. 1 (2006): 112–126; Patrick Devine Wright, "Rethinking NIMBYism: The Role of Place Attachment and Place Identity in Explaining Place-Protective Action," *Journal of Community and Applied Social Psychology* 19, no. 6 (2009): 426–441.

8. Derrick Jensen, "Beyond Hope," *Orion Magazine*, May/June 2006.

CHAPTER 7: COVETED

1. Lauren E. Oakes, Paul E. Hennon, Kevin L. O'Hara, and Rodolfo Dirzo, "Long-Term Vegetation Changes in a Temperate Forest Impacted by Climate Change," *Ecosphere* 5, no. 10 (2014): 1–28.

2. Parker E. Calkin, "Holocene Glaciation of Alaska (and Adjoining Yukon Territory, Canada)," *Quaternary Science Reviews* 7, no. 2 (1988): 159–184; David J. Barclay, Gregory C. Wiles, and Parker E. Calkin, "Holocene Glacier Fluctuations in Alaska," *Quaternary Science Reviews* 28, no. 21 (2009): 2034–2048; Olga N. Solomina, Raymond S. Bradley, Vincent Jomelli, Aslaug Geirsdottir, Darrell S. Kaufman, Johannes Koch, Nicholas P. McKay, et al., "Glacier Fluctuations During the Past 2000 Years," *Quaternary Science Reviews* 149 (2016): 61–90.

3. Paul E. Hennon, David V. D'Amore, Paul G. Schaberg, Dustin T. Wittwer, and Colin S. Shanley, "Shifting Climate, Altered Niche, and a Dynamic Conservation Strategy for Yellow-Cedar in the North Pacific Coastal Rainforest," *BioScience* 62, no. 2 (2012): 147–158.

4. Daniel Stokols and Sally A. Shumaker, "People in Places: A Transactional View of Settings," *Cognition, Social Behavior, and the Environment* (1981): 441–488; Daniel R. Williams and Joseph W. Roggenbuck, "Measuring Place Attachment: Some Preliminary Results," paper presented at NRPA Symposium on Leisure Research, San Antonio, Texas, 1989; Daniel R. Williams and Jerry J. Vaske, "The Measurement of Place Attachment: Validity and Generalizability of a Psychometric Approach," *Forest Science* 49, no. 6 (2003): 830–840; Elisabeth Kals, Daniel Schumacher, and Leo Montada, "Emotional Affinity Toward Nature as a Motivational Basis to Protect Nature," *Environment and Behavior* 31, no. 2 (1999): 178–202; Ruth Rogan, Moira O'Connor, and Pierre Horwitz, "Nowhere to Hide: Awareness and Perceptions of Environmental Change, and Their Influence on Relationships with Place," *Journal of Environmental Psychology* 25, no. 2 (2005): 147–158; Leila Scannell and Robert Gifford, "Defining Place Attachment: A Tripartite Organizing Framework," *Journal of Environmental Psychology* 30, no. 1 (2010): 1–10.

5. Thomas J. Thornton, ed., *Haa Leelk'w Has Aani Saax'u / Our Grandparents' Names on the Land* (Seattle: University of Washington Press, 2012).

6. Wayne W. Leighty, Steven P. Hamburg, and John Caouette, "Effects of Management on Carbon Sequestration in Forest Biomass in Southeast Alaska," *Ecosystems* 9, no. 7 (2006): 1051–1065.

7. Joseph A. Roos, Daisuke Sasatani, Allen M. Brackley, and Valerie Barber, "Recent Trends in the Asian Forest Products Trade and Their Impact on Alaska," US Department of Agriculture, Forest Service, Pacific Northwest Research Station, PNW-RN-564, 2010, 1–42; "Log Value Up," Masthead at Press Time, *Maritime Digest*, June 9, 1973, 2.

8. Inga Petaisto, regional valuation forester, R10, United States Forest Service, email message to author, November 2, 2017; data summarized by Sawa Francis, data clerk, R10, United States Forest Service, from Historic Timber Cut-Sold Reports, available at https://www.fs.fed.us/forestmanagement/products/cut-sold/index.shtml.

9. Donald B. Zobel, "*Chamaecyparis* Forests," in *Coastally Restricted Forests*, edited by Aimlee D. Laderman (Oxford: Oxford University Press on Demand, 1998), 39–53.

10. Paula Dobbyn, "Icy Straits Lumber and Milling Co. Expands in SE," *Alaska Journal of Commerce*, August 9, 2012, www.alaskajournal.com/community/2012-08-09/icy-straits-lumber-and-milling-co-expands-se.

CHAPTER 8: APART AND A PART

1. Thomas F. Thornton, *Being and Place Among the Tlingit* (Seattle: University of Washington Press, 2011), 43–44.

2. Madonna L. Moss, "The Status of Archaeology and Archaeological Practice in Southeast Alaska in Relation to the Larger Northwest Coast," *Arctic Anthropology* 41, no. 2 (2004): 177–196; John Lindo, Alessandro Achilli, Ugo A. Perego, David Archer, Cristina Valdiosera, Barbara Petzelt, Joycelynn Mitchell, et al., "Ancient Individuals from the North American Northwest Coast Reveal 10,000 Years of Regional Genetic Continuity," *Proceedings of the National Academy of Sciences* 114, no. 16 (2017): 4093–4098.

3. Madonna Moss, Justin M. Hays, Peter M. Bowers, and Douglas Reger, "The Archaeology of Coffman Cove, 5500 Years of Settlement in the Heart of Southeast Alaska," University of Oregon Anthropological Papers, 2016, No. 72, 17.

4. "Eyak, Tlingit, Haida, and Tsimshian Cultures of Alaska," Alaska Native Heritage Center, n.d., www.alaskanative.net/en/main-nav/education-and-programs/cultures-of-alaska/eyak-tlingit-haida-and-tsimshian.

5. Paul E. Hennon, Carol M. McKenzie, David D. D'Amore, Dustin T. Wittwer, Robin L. Mulvey, Melinda S. Lamb, Frances E. Biles, and Rich C. Cronn, "A Climate Adaptation Strategy for Conservation and Management of Yellow-Cedar in Alaska," US Department of Agriculture, Forest Service, Pacific Northwest Research Station, PNW-GTR-917, January 2016, 2–4; Alden Springer Crafts and Carl E. Crisp, *Phloem Transport in Plants* (San Francisco: W. H. Freeman, 1971).

6. Nancy J. Turner, "Plants in British Columbia Indian Technology," Handbook no. 38 (Victoria: Royal British Columbia Museum, 1979), 309.

7. Nancy Turner, *Ancient Pathways, Ancestral Knowledge: Ethnobotany and Ecological Wisdom of Indigenous Peoples of Northwestern North America*, vol. 1 (Montreal: McGill-Queen's University Press, 2014), 27.

8. Wallace Stegner to David E. Pesonen, December 3, 1960, in Wallace Stegner, *The Sound of Mountain Water* (New York: Doubleday, 1969), 145–153, reproduced at "Wallace Stegner," Wilderness Society, https://wilderness.org/bios/former-council-members/wallace-stegner.

9. Philip L. Fradkin, *Wallace Stegner and the American West* (Berkeley: University of California Press, 2009).

10. Wilderness Act, Pub. L. No. 88-577, U.S.C. § 1(c), 1131–1136 (1964).

11. Stegner to Pesonen, in Stegner, *Sound of Mountain Water*.

12. Roderick Nash, *Wilderness and the American Mind*, 3rd ed. (New Haven, CT: Yale University Press, 1982), and Roderick Nash, "Wilderness Is All in Your Mind," *Backpacker 31* 7, no. 1 (February/March 1979): 39–41, 70–75.

13. William Cronon, "The Trouble with Wilderness: Or, Getting Back to the Wrong Nature," *Environmental History* 1, no. 1 (1996): 7–28; William Cronon, ed., "The Trouble with Wilderness: Or, Getting Back to the Wrong Nature," in William Cronon, *Uncommon Ground: Rethinking the Human Place in Nature* (New York: W. W. Norton, 1995), 69–90.

14. Story adapted from the original told by Alice Paul in Hesquiat, with permission: Nancy J. Turner and Barbara S. Efrat, *Ethnobotany of the Hesquiat Indians of Vancouver Island*, Cultural Recovery Paper No. 2 (Victoria: British Columbia Provincial Museum, 1982), 33.

15. Cyrus E. Peck, *The Tides People: Tlingit Indians of Southeast Alaska* (C. E. Peck, 1986).

16. Kari M. Norgaard, *Living in Denial: Climate Change, Emotions, and Everyday Life* (Cambridge, MA: MIT Press, 2011), xix, 8.

17. Paul C. Stern and Thomas Dietz, "The Value Basis of Environmental Concern," *Journal of Social Issues* 50, no. 3 (1994): 65–84.

18. P. Wesley Schultz, "The Structure of Environmental Concern: Concern for Self, Other People, and the Biosphere," *Journal of Environmental Psychology* 21, no. 4 (2001): 327–339.

CHAPTER 9: SATURATION POINT

1. R. Bryan Kennedy and D. Ashley Kennedy, "Using the Myers-Briggs Type Indicator® in Career Counseling," *Journal of Employment Counseling* 41, no. 1 (2004): 38–43.

2. Carl G. Jung, *Collected Works of CG Jung*, vol. 6, *Psychological Types*, edited by Gerhard Adler and R. F. C. Hull (Princeton, NJ: Princeton University Press, 1971), 169–170.

3. Isabel B. Myers and Peter B. Myers, *Gifts Differing* (Palo Alto, CA: Consulting Psychologists Press, 1980).

CHAPTER 10: MEASURED AND IMMEASURABLE

1. Center for Biological Diversity, The Boat Company, Greater Southeast Alaska Conservation Community, and Greenpeace, "Petition to List

Yellow-Cedar, *Callitropsis nootkatensis*, Under the Endangered Species Act," Notice of Petition to Sally Jewell, Secretary of the Interior, US Department of the Interior, June 24, 2014, 59.

2. Ibid, 8.

3. Sandra Díaz and Marcelo Cabido, "Vive la Différence: Plant Functional Diversity Matters to Ecosystem Processes," *Trends in Ecology and Evolution* 16, no. 11 (2001): 646–655.

4. Bill McKibben, *Oil and Honey: The Education of an Unlikely Activist* (New York: St. Martin's Press, 2013).

5. David L. Uzzell, "The Psycho-Spatial Dimension of Global Environmental Problems," *Journal of Environmental Psychology* 20, no. 4 (2000): 307–318.

6. James A. Hansen, "The Greenhouse Effect: Impacts on Current Global Temperature and Regional Heat Waves," presented to the Committee on Energy and Natural Resources, US Senate, Washington, DC, June 23, 1988.

7. Richard A. Kerr, "Hansen vs. the World on the Greenhouse Threat," *Science* 244, no. 4908 (1989): 1041–1044.

8. James Hansen and Sergej Lebedeff, "Global Trends of Measured Surface Air Temperature," *Journal of Geophysical Research: Atmospheres* 92, no. D11 (1987): 13345–13372; James Hansen and Sergej Lebedeff, "Global Surface Air Temperatures: Update Through 1987," *Geophysical Research Letters* 15, no. 4 (1988): 323–326.

9. John Tyndall, "On the Absorption and Radiation of Heat by Gases and Vapours, and on the Physical Connection of Radiation, Absorption, and Conduction," *Philosophical Transactions of the Royal Society of London* 151 (1861): 1–36.

10. Svante Arrhenius, "On the Influence of Carbonic Acid in the Air upon the Temperature of the Ground," *Philosophical Magazine and Journal of Science*, series 5, vol. 41 (1896): 237–276.

11. Mark Bowen, *Censoring Science: Inside the Political Attack on Dr. James Hansen and the Truth About Global Warming* (New York: Penguin, 2008).

12. Bill McKibben, *The End of Nature* (New York: Random House, 1989), 51.

13. McKibben, *Oil and Honey*, 108.

14. Anthony Leiserowitz, Edward Maibach, Connie Roser-Renouf, Geoffrey Feinberg, and Seth Rosenthal, *Global Warming's Six Americas*, Yale University and George Mason University (New Haven, CT: Yale Program on Climate Change Communication, 2015). The numbers I reference are from the update to the original 2009 study; see http://climatecommunication.yale.edu/visualizations-data/six-americas.

15. Connie Roser-Renouf, Edward Maibach, Anthony Leiserowitz, and Seth Rosenthal, *Global Warming's Six Americas and the Election*, Yale University

and George Mason University (New Haven, CT: Yale Program on Climate Change Communication, 2016).

16. Michael E. Mann, *The Hockey Stick and the Climate Wars: Dispatches from the Front Lines* (New York: Columbia University Press, 2012), 253.

17. John H. Richardson, "When the End of Human Civilization Is Your Day Job," *Esquire*, July 7, 2015, www.esquire.com/news-politics/a36228/ballad-of-the-sad-climatologists-0815.

18. Rebecca Solnit, *Hope in the Dark: Untold Histories, Wild Possibilities*, 3rd ed. (Chicago: Haymark Books, 2016), 22.

19. Naomi Klein, *This Changes Everything: Capitalism vs. the Climate* (New York: Simon and Schuster, 2014), 21.

CHAPTER 11: THE GREATEST OPPORTUNITY

1. Nathan G. McDowell, A. P. Williams, C. Xu, W. T. Pockman, L. T. Dickman, S. Sevanto, R. Pangle, et al., "Multi-Scale Predictions of Massive Conifer Mortality Due to Chronic Temperature Rise," *Nature Climate Change* 6, no. 3 (2016): 295–300; Craig D. Allen, "Forest Ecosystem Reorganization Underway in the Southwestern United States: A Preview of Widespread Forest Changes in the Anthropocene?," in *Forest Conservation in the Anthropocene: Science, Policy, and Practice*, edited by Sample V. Alaric, Bixler R. Patrick, and Miller Char (Boulder: University Press of Colorado, 2016), 57–70; Craig D. Allen, David D. Breshears, and Nate G. McDowell, "On Underestimation of Global Vulnerability to Tree Mortality and Forest Die-Off from Hotter Drought in the Anthropocene," *Ecosphere* 6, no. 8 (2015): 1–55; Amy C. Bennett, Nathan G. McDowell, Craig D. Allen, and Kristina J. Anderson-Teixeira, "Larger Trees Suffer Most During Drought in Forests Worldwide," *Nature Plants* 1, no. 10 (2015): 15139.

2. Cally Carswell, "The Tree Coroners: To Save the West's Forests, Scientists Must First Learn How Trees Die," *High Country News*, December 9, 2013.

3. Michelle Nijhuis, "For the Love of Trees," *High Country News*, December 9, 2013.

4. Craig D. Allen, "Changes in the Landscapes of the Jemez Mountains, New Mexico" (PhD diss., University of California–Berkeley, 1989), 253–254.

5. Craig D. Allen and David D. Breshears, "Drought-Induced Shift of a Forest-Woodland Ecotone: Rapid Landscape Response to Climate Variation," *Proceedings of the National Academy of Sciences* 95, no. 25 (1998): 14839–14842.

6. William R. L. Anderegg, Jeffrey M. Kane, and Leander D. L. Anderegg, "Consequences of Widespread Tree Mortality Triggered by Drought

and Temperature Stress," *Nature Climate Change* 3, no. 1 (2013): 30–36; William R. L. Anderegg, Lenka Plavcová, Leander D. L. Anderegg, Uwe G. Hacke, Joseph A. Berry, and Christopher B. Field, "Drought's Legacy: Multiyear Hydraulic Deterioration Underlies Widespread Aspen Forest Die-Off and Portends Increased Future Risk," *Global Change Biology* 19, no. 4 (2013): 1188–1196; William R. L. Anderegg, Joseph A. Berry, Duncan D. Smith, John S. Sperry, Leander D. L. Anderegg, and Christopher B. Field, "The Roles of Hydraulic and Carbon Stress in a Widespread Climate-Induced Forest Die-Off," *Proceedings of the National Academy of Sciences* 109, no. 1 (2012): 233–237.

7. William R. L. Anderegg, "Good Night, Sweet Trees: Sudden Aspen Decline Is Like a Shakespearean Tragedy," *High Country News*, February 26, 2010.

8. Eric Klinenberg, "Adaptation: How Can Cities Be 'Climate-Proofed'?" *The New Yorker*, January 7, 2013.

CHAPTER 12: THE SENTINELS

1. Rainer Maria Rilke, *Letters to a Young Poet*, translated by M. D. Herter Norton, rev. ed. (New York: W. W. Norton, 2004 [1934]).

2. Gregory C. Wiles, Daniel E. Lawson, Nick Wiesenberg, Caitlin Fetters, and Brian Tracy, "Tree Ring Dating, Glacier Dynamics, and Tlingit Ethnographic Histories of Little Ice Age Environmental Change in Glacier Bay National Park and Preserve, Alaska," manuscript in prep.

3. Mary Beth Moss, email messages to author, November 13, 2017, and May 11, 2018.

4. Sarah McGrath, Greg Wiles, Wayne Howell, Nick Wiesenberg, and Colin Mennett, "Tree Ring Dating of Traditional Native Bark Stripping on Pleasant Island, Icy Strait, Southeast Alaska, USA," manuscript in prep.

5. Gay Alcorn, "Tony Abbott and Naomi Klein Agree: We Can't Beat Climate Change Under Capitalism," *The Guardian*, September 3, 2015, https://www.theguardian.com/commentisfree/2015/sep/03/tony-abbott-and-naomi-klein-agree-we-cant-beat-climate-change-under-capitalism; Naomi Klein, *This Changes Everything: Capitalism vs. the Climate* (New York: Simon and Schuster, 2014), 452.

EPILOGUE

1. John Krapek, Paul E. Hennon, David V. D'Amore, and Brian Buma, "Despite Available Habitat at Range Edge, Yellow-Cedar Migration Is Punctuated with a Past Pulse Tied to Colder Conditions," *Diversity and Distributions* (2017): 1–12.

2. John Krapek and Brian Buma, "Limited Stand Expansion by a Long-Lived Conifer at a Leading Northern Range Edge, Despite Available Habitat," *Journal of Ecology* 106, no. 3 (2018): 911–924.

3. Maria L. La Ganga, "Alaska Yellow Cedar Closer to Endangered Species Act Protection," *Los Angeles Times*, April 10, 2015, www.latimes.com/nation/nationnow/la-na-nn-alaska-yellow-cedar-20150410-story.html.

4. Paul E. Hennon, Carol M. McKenzie, David V. D'Amore, Dustin T. Wittwer, Robin L. Mulvey, Melinda S. Lamb, Frances F. Biles, and Rich C. Cronn, "A Climate Adaptation Strategy for Conservation and Management of Yellow-Cedar in Alaska," US Department of Agriculture, Forest Service, Pacific Northwest Research Station, PNW-GTR-917, January 2016.

5. Brian Buma, Paul E. Hennon, Constance A. Harrington, Jamie R. Popkin, John Krapek, Melinda S. Lamb, Lauren E. Oakes, Sari Saunders, and Stefan Zeglen, "Crossing the Snow-Rain Threshold: Emerging Mortality over 10 Degrees of Latitude of a Climate-Threatened Conifer," *Global Change Biology* 23, no. 7 (2017): 2903–2914.

6. Tara M. Barrett and Robert R. Pattison, "No Evidence of Recent (1995–2013) Decrease of Yellow-Cedar in Alaska," *Canadian Journal of Forest Research* 47, no. 999 (2017): 97–105.

7. Brian Buma, "Transitional Climate Mortality: Slower Warming May Result in Increased Climate-Induced Mortality in Some Systems," *Ecosphere* 9, no. 3 (2018): e02170.

8. Alison Bidlack, Sarah Bisbing, Brian Buma, David V. D'Amore, Paul E. Hennon, Thomas Heutte, John Krapek, Robin Mulvey, and Lauren E. Oakes, "Alternative Interpretation and Scale-Based Context for 'No Evidence of Recent (1995–2013) Decrease in Yellow-Cedar in Alaska' (Barrett and Pattison 2017)," *Canadian Journal of Forest Research* 47, no. 8 (2017): 1145–1151.

Further Reading

The study that resulted from John Caouette's collaboration with E. Ashley Steel:

Caouette, J. P., E. A. Steel, P. E. Hennon, P. G. Cunningham, C. A. Pohl, and B. A. Schrader. "Influence of Elevation and Site Productivity on Conifer Distributions Across Alaskan Temperate Rainforests." *Canadian Journal of Forest Research* 46, no. 2 (2015): 249–261.

For historical climate information and the distribution of forests:

Davis, Margaret B. "Quaternary History of Deciduous Forests of Eastern North America and Europe." *Annals of the Missouri Botanical Garden* 70, no. 3 (1983): 550–563.

Davis, Margaret B., and Ruth G. Shaw. "Range Shifts and Adaptive Responses to Quaternary Climate Change." *Science* 292, no. 5517 (2001): 673–679.

Graham, Russel W., and Eric C. Grimm. "Effects of Global Climate Change on the Patterns of Terrestrial Biological Communities." *Trends in Ecology and Evolution* 5, no. 9 (1990): 289–292.

Webb, Thompson III, and Patrick J. Bartlein. "Global Changes During the Last 3 Million Years: Climatic Controls and Biotic Responses." *Annual Review of Ecology and Systematics* 23, no. 11 (1992): 141–173.

On permafrost and carbon in the North:

Hollesen, Jørgen, Henning Matthiesen, Anders Bjørn Møller, and Bo Elberling. "Permafrost Thawing in Organic Arctic Soils Accelerated by Ground Heat Production." *Nature Climate Change* 5, no. 6 (2015): 574–578.

Hugelius, Gustaf, Tarmo Virtanen, Dmitry Kaverin, Alexander Pastukhov, Felix Rivkin, Sergey Marchenko, Peter Kuhry, et al. "High-Resolution Mapping of Ecosystem Carbon Storage and Potential Effects of Permafrost Thaw in Periglacial Terrain, European Russian Arctic." *Journal of Geophysical Research: Biogeosciences* 116, no. G3 (2011).

Schuur, E. A. G., A. D. McGuire, C. Schädel, G. Grosse, J. W. Harden, D. J. Hayes, G. Hugelius, et al. "Climate Change and the Permafrost Carbon Feedback." *Nature* 520, no. 7546 (2015): 171–179.

On forest dieback events and plant communities in a changing climate:

Anderegg, William R. L., Joseph A. Berry, Duncan D. Smith, John S. Sperry, Leander D. L. Anderegg, and Christopher B. Field. "The Roles of Hydraulic and Carbon Stress in a Widespread Climate-Induced Forest Die-Off." *Proceedings of the National Academy of Sciences* 109, no. 1 (2012): 233–237.

Anderegg, William R. L., Jeffrey M. Kane, and Leander D. L. Anderegg. "Consequences of Widespread Tree Mortality Triggered by Drought and Temperature Stress." *Nature Climate Change* 3, no. 1 (2013): 30–36.

Hansen, Andrew J., Ronald P. Neilson, Virginia H. Dale, Curtis H. Flather, Louis R. Iverson, David J. Currie, Sarah Shafer, Rosamonde Cook, and Patrick J. Bartlein. "Global Change in Forests: Responses of Species, Communities, and Biomes: Interactions Between Climate Change and Land Use Are Projected to Cause Large Shifts in Biodiversity." *BioScience* 51, no. 9 (2001): 765–779.

McDowell, Nate, William T. Pockman, Craig D. Allen, David D. Breshears, Neil Cobb, Thomas Kolb, Jennifer Plaut, et al. "Mechanisms of Plant Survival and Mortality During Drought: Why Do Some Plants Survive While Others Succumb to Drought?" *New Phytologist* 178, no. 4 (2008): 719–739.

Sevanto, Sanna, Nate G. McDowell, L. Turin Dickman, Robert Pangle, and William T. Pockman. "How Do Trees Die? A Test of the Hydraulic Failure and Carbon Starvation Hypotheses." *Plant, Cell & Environment* 37, no. 1 (2014): 153–161.

Walther, Gian-Reto, Eric Post, Peter Convey, Annette Menzel, Camille Parmesan, Trevor J. C. Beebee, Jean-Marc Fromentin, Ove Hoegh-Guldberg, and Franz Bairlein. "Ecological Responses to Recent Climate Change." *Nature* 416, no. 6879 (2002): 389–395.

On the botanical debate around the genus:

Farjon, A., Nguyen Tien Hiep, D. K. Harder, Phan Ke Loc, and L. Averyanov. "A New Genus and Species in Cupressaceae (Coniferales) from Northern Vietnam, *Xanthocyparis vietnamensis*." *Novon* 12, no. 2 (2002): 179–189.

Oersted, Anders S. "Bidrag til Naaletraeernes Morphologi, Videnskabelige Meddelelser fra Dansk Naturhistorisk Forening I Kjobenhavn" [Contributions to the Morphology of Conifers, Contributions to the Natural History Society of Copenhagen], Series 2, no. 6 (1864): 1–36.

Spach, Édouard. *Histoire Naturelle des Végétaux. Phanérogames*. Paris: Librairie Encyclopédique De Roret, 1842.

First residents of the Northwest Coast:

Carlson, Roy L. "Trade and Exchange in Prehistoric British Columbia." In *Prehistoric Exchange Systems in North America*, edited by T. G. Baugh and J. E. Ericson, 307–361. London: Springer, 1994.

Davis, Stanley D. "Prehistory of Southeastern Alaska." In *Handbook of North American Indians*. Vol. 7, *Northwest Coast*, edited by Wayne Suttles and William C. Sturtevant, 197–202. Washington, DC: Smithsonian Institute, 1990.

Dixon, E. James. *Bones, Boats, and Bison*. Albuquerque: University of New Mexico Press, 1999.

Matson, R. G., and Cary Coupland. *The Prehistory of the Northwest Coast*. San Diego: Academic Press, 1995.

Moss, Madonna L. *Northwest Coast: Archaeology as Deep History*. Washington, DC: SAA Press, 2011.

Moss, Madonna L., and Jon M. Erlandson. "Reflections on North American Pacific Coast Prehistory." *Journal of World Prehistory* 9, no. 1 (1995): 1–45.

Index

Credit: Clayton Boyd

Lauren E. Oakes is a conservation scientist at the Wildlife Conservation Society and an adjunct professor in Earth System Science at Stanford University. She lives in Portola Valley, California, and Bozeman, Montana.